I0043715

Joseph Anthony Gillet, William Jame Rolfe

Elements of natural philosophy :

for the Use of Schools and Academies

Joseph Anthony Gillet, William Jame Rolfe

Elements of natural philosophy :
for the Use of Schools and Academies

ISBN/EAN: 9783337025045

Printed in Europe, USA, Canada, Australia, Japan

Cover: Foto ©berggeist007 / pixelio.de

More available books at **www.hansebooks.com**

ELEMENTS

OF

NATURAL PHILOSOPHY

FOR THE USE OF

Schools and Academies.

BY

J. A. GILLET,

PROFESSOR OF PHYSICS IN THE NORMAL COLLEGE OF THE CITY OF NEW YORK,

AND

W. J. ROLFE,

FORMERLY HEAD MASTER OF THE HIGH SCHOOL,
CAMBRIDGE, MASS.

POTTER, AINSWORTH, & CO.
NEW YORK AND CHICAGO.
1884.

PREFACE.

——◆——

THIS book is an abridgment of the "Natural Philosophy" by the same authors, with such changes as were required to adapt it to younger pupils. As the larger book is in no sense a revision of the "Natural Philosophy" of the "Cambridge Course of Physics," but an entirely new work, differing from its predecessor both in matter and in method of presentation, so this book is equally independent of the earlier "Handbook of Natural Philosophy."

Two kinds of type have been used, as in the larger book, with a view to adapting it to the wants of different schools. The matter in the larger type forms a brief and easy course, complete in itself and sufficient for classes that can spend only a single school term in the study. The portions in smaller type will enable teachers to extend this course more or less, according to their personal tastes or the ability of their pupils. Those who have time for the whole book may find the division convenient in reviews and examinations, and also in fixing the minimum to be required of pupils who have little taste or aptitude for physics.

The authors have carefully avoided doing all the teacher's work for him by anticipating every familiar illustration which he would either give his pupils orally or lead them to see and state for themselves. Teachers may be supposed to know something, and it is a very dull pupil that knows nothing at all, except what is "in the book." Both teacher and pupil become mere machines when the one has only to hear the other repeat what he has learned by rote from the printed page. Such practical hints and suggestions as may be needed by the young and inexperienced instructor had better be furnished him outside of the text-book.

It may be added that the book of which this is an abridgment will be useful in various ways to the teacher of this; and he will also get much help from any of the following works, to which the authors have already acknowledged their indebtedness in the preface to the larger book : —

Deschanel's Natural Philosophy. D. Appleton & Co.: New York (reprint).

Ganot's Physics. Wm. Wood & Co.: New York (reprint).

Tait's Recent Advances in Physical Science. Macmillan & Co.: New York.

Maxwell's Matter and Motion. Macmillan & Co. : New York.

Tyndall's Sound. D. Appleton & Co. : New York (reprint).

Mayer's Sound. D. Appleton & Co. : New York.

Helmholtz's Popular Lectures. 1st Series. D. Appleton & Co. : New York (reprint).

Taylor's Sound and Music. Macmillan & Co. : New York.

Tyndall's Heat a Mode of Motion. D. Appleton & Co. : New York (reprint).

Maxwell's Theory of Heat. D. Appleton & Co.: New York (reprint).

Mayer's Light. D. Appleton & Co. : New York.

Tyndall's Lectures on Light.   D. Appleton & Co.: New York.
Rood's Modern Chromatics.   D. Appleton & Co.: New York.
Jeffries's Color Blindness.   Houghton, Mifflin, & Co.: Boston.
Gordon's Electricity and Magnetism.   D. Appleton & Co.:
New York (reprint).
Jenkin's Electricity and Magnetism.   D. Appleton & Co.;
New York (reprint).
Tyndall's Lessons in Electricity.   D. Appleton & Co.: New
York (reprint).
Prescott's Telegraph.   D. Appleton & Co.: New York.
Prescott's Telephone, etc.   D. Appleton & Co.: New York.
Sawyer's Electric Lighting.   D. Van Nostrand: New York.
Loomis's Meteorology.   Harper & Brothers: New York.
Stewart's Energy.   D. Appleton & Co.: New York.

CONTENTS.

———

ELEMENTS

OF

NATURAL PHILOSOPHY.

ELEMENTS

OF

NATURAL PHILOSOPHY.

————◆————

I.

CONSTITUTION OF MATTER.

1. *Molecules and Atoms.* — All bodies are supposed to be made up of very small particles, called *molecules*, which are in turn made up of still smaller particles, called *atoms*.

These molecules are far too minute to be seen with the most powerful microscope, and are separated by spaces many times as large as the molecules themselves. It has been estimated that there are at least 300 quintillions of molecules in one cubic inch of air, — a number which would be represented by 3 followed by twenty ciphers. At the same time it is believed that the material molecules themselves occupy only $\frac{1}{3000}$ of the space in the cubic inch. The atoms that make up the molecules are also believed to be very far apart compared with their size. We thus gain some notion of the extreme fineness of the atomic dust of which matter is composed.

We can resolve bodies into molecules, and molecules into atoms; but it has not been found possible to divide the atoms.

2. *Substance.* — The *substance* of a body depends upon *the internal structure of its molecules*. All the molecules of the same substance are supposed to be exactly alike. A

I

body may be divided and subdivided at will, and the substance of every portion will remain the same so long as the molecules are unchanged. If the molecules are divided, or their structure is altered by changing the kind, number, or grouping of their atoms, the substance of the body is changed.

If a piece of iron is reduced to the finest powder, every particle is iron still; but if the iron *rusts*, its molecules unite with those of oxygen in the air, forming more complex molecules and a new substance. Ice may be changed to water, and water to steam, without change of substance, for the molecules remain the same; but if we divide these molecules by chemical processes we obtain oxygen and hydrogen, two substances made up of less complex molecules.

3. *The Ether.* — A highly rarefied fluid, called the *ether*, is supposed to *fill all space* and to *permeate all bodies*. It fills alike the spaces among the planets and stars and those among molecules and atoms. It is without weight, and offers no resistance to bodies, molecules, or atoms moving about in it.

4. *The Structure of Bodies analogous to that of the Sidereal Universe.* — The Sidereal Universe is composed of stars, each of which is probably, like our own sun, the centre of a solar system composed of sun and planets. The *planets and moons* which compose a solar system correspond to the *atoms* which compose the molecules, and the *solar systems* correspond to the *molecules* which compose the body. The planets in the solar system are sometimes found singly, as in the case of Venus, and sometimes in groups, as in the case of Jupiter and his moons. The same is true of the atoms in the molecules.

5. *All Matter is Porous.* — From what has been said, it will be evident that all matter is *porous*, that is, it contains spaces which are *not occupied with material particles*. When these pores are too small to be seen with the microscope, they are called *physical* pores. In wood and many other

substances the pores are large enough to be seen; they are then called *sensible* pores.

6. *The Three Orders of Material Units.* — The three orders of material units are *atoms, molecules,* and *bodies.*

7. *Atomic, Molecular, and Molar Motion.* — Every particle of matter in the universe is in *incessant motion.* The atoms are all the time moving about in the molecules; the molecules, in bodies; and bodies, in space. The motion of the *atoms* within the molecules is called *atomic* motion; that of the *molecules* in bodies, *molecular* motion; and that of *bodies* in space, *molar* motion. Molar motion is often called *mechanical* motion. Sometimes the term *molecular* is applied to the motion of both atoms and molecules.

8. *The Three Great Forces of Nature.* — There are *three forces* corresponding to the three orders of material units. These are *affinity, cohesion,* and *gravity.*

Affinity is the force which *binds together the atoms into molecules.* It is therefore an *atomic* force. It is the strongest of the forces, but it acts only through infinitesimal distances.

Cohesion is a *molecular* force. It *binds together the molecules into bodies.* It is a weaker force than affinity, but is capable of acting through greater, though still insensible distances.

Gravity is a *molar* force. It *binds together bodies.* It is the weakest of the three forces, but is capable of acting through all known distances.

9. *Elasticity.* — Elasticity is the *tendency of a body to spring back to its original condition* when it has been distorted in any way.

Any *distortion*, whether produced by stretching, by bending, by twisting, by compression, or by rarefaction, is called a *strain.* The force which produces the strain is called a *stress.* Elasticity is always developed by some kind of strain. All bodies are elastic to some extent, but usually, when the distortion proceeds beyond a certain point, the elasticity of the body breaks down. This point is called the *limit* of the elasticity of the body.

10. *Chemical Properties of Matter.* — The properties of matter which grow out of the atomic structure of the molecules and the action of affinity are called *chemical properties*.

11. *Physical Properties of Matter.* — The properties of matter which grow out of the molecular structure of bodies and the action of cohesion are called *physical properties*.

12. *The Physical Sciences.* — The *physical sciences* deal with the action of forces on material units, irrespective of the phenomena of life.

Mechanics deals with the action of forces and the laws of motion, irrespective of any order of material units.

Astronomy deals with gravity and molar units.

Physics deals with cohesion, molecules, and physical properties of matter.

Chemistry deals with affinity, atoms, and chemical properties of matter.

Natural Philosophy includes both Mechanics and Physics.

II.

MECHANICS.

A. Definitions.— Units. — Newton's Laws of Motion.

13. *The Three Fundamental Units.* — The three *fundamental* units of Mechanics, from which all the other mechanical and physical units are derived, are the *unit of time*, the *unit of length*, and the *unit of mass*.

In the English system these units are the *second*, the *foot*, and the *pound* (avoirdupois). In the French system they are the *second*, the *centimetre*, and the *gramme*.

14. *English and French Units of Length.* — The English standard unit of length is the *yard*, which is divided into three equal parts, called *feet*. The foot is subdivided into twelve equal parts, called *inches*. The yard is simply the length marked on a certain rod preserved by the government.

The French standard unit of length is the *metre*. This is, theoretically, the ten-millionth of the distance from the equator to the pole, or the forty-millionth of the distance round the earth. Practically, it is the length of a rod preserved by the French government, which differs appreciably from the theoretical length of the metre. The metre is about $3\frac{1}{4}$ feet. It is divided into ten, one hundred, and one thousand equal parts, called *decimetres, centimetres,* and *millimetres. Decametre, hectometre,* and *kilometre* are, respectively, ten metres, one hundred metres, and one thousand metres. In the French or *Metric* system of units the prefixes *deci, centi,* and *milli* always indicate tenths, hundredths, and thousandths of the unit, while the prefixes *deca*, *hecto*, and *kilo* always indicate tens, hundreds, and thousands of the units.

For readily comparing the French units of length with our familiar English units, it will be convenient to remember that a metre is about forty inches; a decimetre, about four inches; a centimetre about $\frac{4}{10}$ of an inch; and a millimetre, about $\frac{1}{25}$ of an inch. A kilometre is about five furlongs, or $\frac{5}{8}$ of a mile.

15. *Units of Surface and of Volume.* — The units of surface are *squares*, one of whose sides is the unit of length. Thus, the English units of surface are the *square yard*, the *square foot*, and the *square inch*. The French units of surface are the *square metre*, the *square decimetre*, and the *square centimetre*.

The units of volume are *cubes*, one of whose edges is the unit of length. The English units of volume are the *cubic yard*, the *cubic foot*, and the *cubic inch*. The French units of volume are the *cubic metre*, the *cubic decimetre*, and the *cubic centimetre*. The French unit of capacity is the *cubic decimetre*. It is called the *litre*, and is equal to about $1\frac{3}{4}$ pints, or a little less than a quart.

16. *Units of Mass.* — The *mass* of a body is the *quantity of matter* which it contains. The English unit of mass is the mass of a certain piece of metal preserved by the government and called the *pound avoirdupois*. It is divided into 7000 equal parts, called *grains*. The French unit of mass is the mass of a cubic centimetre of water at its maximum density. It is called a *gramme*, and is equal to about $15\frac{1}{2}$ grains. A kilogramme is equal to about $2\frac{1}{5}$ pounds.

17. *Unit of Density.* — The *density* of a body is the *quantity of matter in a unit of its volume*. The density of water at a temperature of 39° F. is usually taken as the unit of density.

18. *Units of Velocity.* — *Velocity* is rate of motion. The

English unit of velocity is the velocity of one *foot a second.* The French unit is the velocity of a *centimetre a second.*

When we speak of the velocity of a body as being five, ten, or twenty feet a second, we mean that, at the instant to which we refer, the body is moving fast enough to go five, ten, or twenty feet in a second, provided it were to keep on moving at the same rate. It does not, however, follow that it will actually go five, ten, or twenty feet in a second, for its rate may change.

19. *The Action of Forces on Matter.* — Any *push* or *pull,* of whatever origin, upon any portion of matter is called a *force.* In the realm of matter these forces always act *between* two different portions of matter. Thus, *affinity* is a pull between two ˙*atoms;* *cohesion,* a pull between two *molecules;* and *gravity,* a pull between two *bodies.*

The action of a pulling, or *attractive,* force may be illustrated by fastening two balls to the ends of an india-rubber cord and then separating the balls so as to stretch the cord. The stretched cord will pull upon both balls. The action of a pushing, or *repulsive,* force may be illustrated by placing a rod of india-rubber between two balls and then crowding the balls together. The compressed rubber will push upon both balls.

This action of a force between two portions of matter takes different names according to the aspect under which it is viewed. When we take into account the whole phenomenon of the action, we call it a *stress.* This stress, according to the mode in which it acts, may be described as *attraction, repulsion, tension, pressure, torsion,* etc. When we confine our attention to one of the portions of matter, we see only one aspect of the stress, namely, that which affects the portion of matter under consideration. This aspect of the phenomenon we call, with reference to its *effect,* an *external force,* acting upon that portion of matter, and, with reference to its *cause,* the *action* of the other portion of matter. The opposite aspect of the stress is called the *reaction* on the other portion of matter.

20. *Newton's First Law of Motion.* — *Every body perseveres in its state of rest or of moving uniformly in a straight line, unless compelled to change this state by external forces,*

This is Newton's first law of motion.   No portion of matter in the universe, so far as known, is absolutely at rest. Were there such a portion of matter, it could be put in motion only by an external force.

Bodies are commonly spoken of as at rest when they are not changing their positions with respect to other bodies around them.   Thus, we say that a body is at rest on the deck of a steamer, though it is really moving forward with the steamer ; and that bodies are at rest on the surface of the earth, though they are moving along with the earth.   In all such cases bodies are only *relatively* at rest.   In common language bodies are said to be at rest with respect to each other when they are all moving along at the same rate and in the same direction. When, in common language, a body is said to be put in motion, what really takes place is that its motion is changed either in rate or direction.

Unless acted upon by external forces, a moving body would *always go on in a straight line and at a uniform rate.* This seems to be contradicted by common experience.   All moving bodies at the surface of the earth show a decided tendency to stop.   But all such moving bodies are acted upon by some external force acting as a *resistance.*   The chief resistances encountered by moving bodies are *friction* and *resistance of the atmosphere.*

In proportion as these resistances are diminished, the longer is the time a body will continue to move.   A smooth stone is soon brought to rest when sliding over the surface of the earth. The same stone will slide much longer over ice, where there is less friction.   A top that will spin for ten minutes in the air will spin more than half an hour in a vacuum.   Since the time a body will continue in motion increases in proportion as the resistance is diminished, we may reasonably infer that, were the resistance entirely removed, the body would continue in motion forever.

21. *Inertia.* — The *tendency of a body to persevere in its state of rest or motion* is called *inertia.*   The inertia of a

body is *directly proportional to its mass.* This inertia must be overcome by some external force in order to put a body in motion, or to change the rate or direction of its motion. It takes *time* for a force to overcome the inertia of matter. Hence, when a body receives a sudden blow, the part of the body immediately receiving the blow yields before there is time to overcome the inertia of the surrounding parts.

There are many striking illustrations of inertia. If a number of checkers are piled up in a column, one of them may be knocked out by a very rapid blow with a table knife without overturning the column. A feeble blow will fail. Stick two needles into the ends of a broomstick and rest the needles on

Fig. 1.

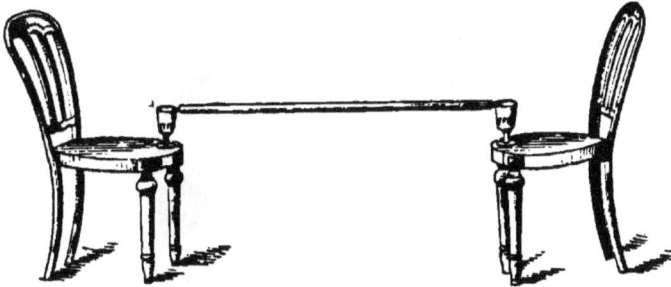

two glass goblets, as shown in Figure 1. Strike the middle of the stick a quick, sharp blow with a heavy poker. The stick will break without breaking the needles or the goblets. Here again a feeble or indecisive blow will fail. A soft body, fired fast enough, will hit as hard as lead. A tallow candle may be fired from a gun through a pine board.

22. *Centrifugal Force.* — The so-called *centrifugal force* is an illustration of Newton's first law of motion. It is simply the *tendency of the parts of a rotating body to keep moving in straight lines.* This tendency increases with the speed of rotation, and sometimes to such a degree as to overcome the cohesion of the body. In this case the body will fly in pieces, as large grindstones and heavy fly-wheels have been known to do. If a stone is fastened to the end

of a string and twirled rapidly around the finger, the tendency of the stone to fly off in a straight line may become sufficient to break the string. In this case the stone will start off in a line tangent to the circle it was describing.

This tendency to move on in a straight line must be counteracted by the force acting towards the centre, in order to keep a body moving in a circle. The faster the body moves, the greater the pull needed to keep the body

Fig. 2.

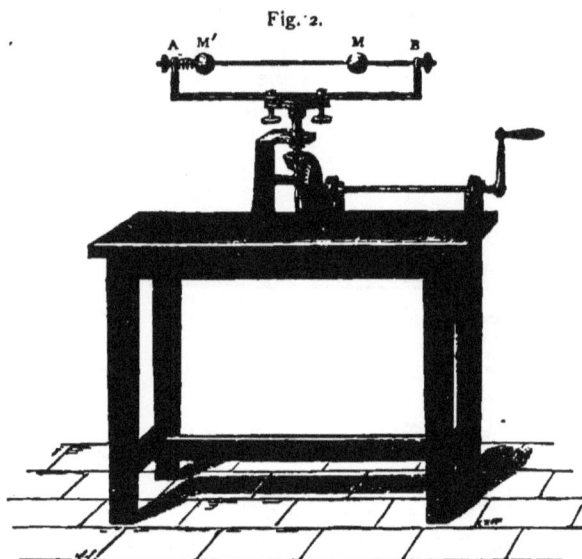

in its circular path. The greater the pull *upon* the body towards the centre, the greater the pull *of* the body away from the centre. The pull upon the body *towards the centre* is called the *centripetal force*, and the pull of the body *away from the centre* is called the *centrifugal force*.

These two forces are only the *two aspects of the stress of attraction between the body and the centre* about which it is revolving.

The pull of a revolving body away from the centre may be illustrated by the pieces of apparatus shown in Figures 2 and 3. In the first, two balls *M* and *M'* are placed on the rod

A B, which passes through them. The rod is then put in rapid rotation by turning the crank, and the balls fly apart.

If the flexible rings of Figure 3 are whirled in place of the rod, they will become more and more flattened as the speed of rotation increases. This change of form is due to the pull of each part of the rings away from the central axis. The pull will be greatest at the central point of the rings, because this part is moving fastest. It was in this way that the earth became flattened at the poles while in the fluid state.

Fig. 3.

The centrifugal railway (Figure 4) shows a curious effect of this outward pull. A carriage starting from *A* descends the incline to *B*, passes up around the circle *C*, and then up the incline to *D*. The outward pull of the carriage due to its velocity is sufficient to keep it against the rails while passing around the circle, though it is part of the time travelling bottom up.

Fig. 4.

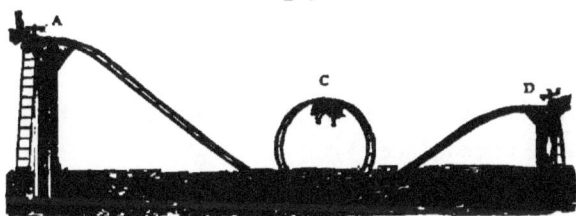

23. *Stability of a Rotating Body.* — The tendency of the particles of matter to keep moving in the same plane explains why a top will stand upright so long as it is spinning rapidly, though it topples over at once as soon as it comes to rest. For the same reason a bicycle is not easily overturned while its large wheel is in rapid rotation.

24. *External Forces tend to put Bodies in Motion or to*

change their Velocities. — Suppose a rubber cord fastened at one end to a body, not acted on by any other force than the tension of the cord, and suppose the cord to be kept stretched to the same extent all of the time, so as to exert a uniform pull upon the body. The body will begin to move in the direction of the pull, and will move faster and faster the longer the pull continues, gaining the same amount of velocity each second. If it were moving at the rate of two feet a second at the end of the first second, it will be moving at the rate of four feet a second at the end of the second second, at the rate of six feet a second at the end of the third second, and so on.

25. *Units of Force.* — Forces may be measured either by *the pressure which they would produce* or by *the rate at which they would increase the velocity of a mass of matter.*

In the former case the *unit of force* is *the force of gravity on a unit of mass.* In the English system it is the force of gravity on the mass of a pound or a grain, and is called a *pound* or a *grain.* In the French system it is the force of gravity on a mass of a gramme, and is called a *gramme.* These units are called *gravitation* units ; and since they depend upon the intensity of gravity, they are *variable,* changing with the intensity of gravity at different places on the surface of the earth, and at different elevations above the surface.

In the latter case the unit of force is *the force that will impart to a unit of mass a unit of velocity in a unit of time.* In the English system it is the force that will impart to a mass of a pound a velocity of a foot in a second. It is called a *poundal.* At Greenwich it takes 32.2 poundals of force to hold up a pound.

A system of absolute measurement has been devised in England, and adopted by the British Association. The units of this system are all based upon the centimetre, gramme, and second as the three fundamental units of length, mass, and time. This system of measurement is called the *centimetre-gramme-second system*, or more briefly, the *C. G. S. system*. Its units are called the *centimetre-gramme-second units*, or more briefly, the *C. G. S. units*.

In the *C. G. S.* system the unit of force is the force that will impart to a mass of a gramme a velocity of one centimetre a second. It is called a *dyne*. It takes 445,000 dynes of force to hold up a pound at Greenwich. These units are independent of gravity, and are *invariable*. They are called *absolute units*.

26. *The Impulse of Force.* — The effect of a force in producing motion is *directly proportional to its intensity and the time during which it acts.* The *product of the intensity and the time during which it acts* is called the *impulse* of the force.

27. *Momentum.* — The motion of a body is measured *by the mass and the velocity of the body*, and is *directly proportional to the two.* If two bodies have equal velocities, but one has five times the mass of the other, it is said to have five times the motion ; or if the two have equal masses, and one has five times the velocity of the other, it is said to have five times the motion of the other. The *product of the mass of a body and its velocity* is called the *momentum* of the body.

28. *Newton's Second Law of Motion.* — *Change of motion is proportional to the impressed force, and takes place in the direction in which the force acts.* This is Newton's second law of motion.

By *motion*, as here used, Newton means what is now called *momentum*, in which the *quantity of matter* moved is taken into account as well as the *rate* at which it travels. For instance, there would be the same change of motion, whether the velocity of four pounds was changed one foot a second, or the velocity of one pound four feet a second. In either case the change of momentum would be four.

By *impressed force* Newton means what is now called *impulse*, in which the *time* the force acts is taken into account as well as the *intensity* of the force. Thus, the impulse, or impressed force, would be the same whether a force of a poundal were acting five seconds or a force of five poundals were acting one second. In either case the impulse, or impressed force, would be five.

Newton's second law, stated in terms of momentum, would be : *The change of momentum of a body is numerically equal to the impulse which produced it, and is in the same direction.*

An unbalanced external force acting upon a body always *changes the velocity of the body in the direction in which it acts.* This change of velocity is called *acceleration.* The acceleration produced in a given time by a force acting upon a body is precisely the same whether the body is at first at rest or in motion, or whether the force is acting alone or with other forces.

When the acceleration is opposed to the original motion of a body, it is usually called a *retardation.*

Newton's second law, stated in terms of acceleration, would be: *When any number of forces act upon a body, the acceleration due to each force is the same in magnitude and direction as if the others had not been in action.*

The total acceleration produced by the action of a force is *directly proportional to the impulse of the force, and inversely proportional to the mass acted upon.* A force of 40 poundals acting for 20 seconds upon a mass of 50 pounds would produce an acceleration of $\overline{40 \times 20} \div 50 = 16$ feet. A force of 300 dynes acting 80 seconds upon 200 grammes would produce an acceleration of $\overline{300 \times 80} \div 200 = 120$ centimetres.

The total change of momentum produced by the action of a force is *numerically equal to the impulse of the force.* A force of 40 poundals acting 30 seconds would produce a change of momentum equal to $40 \times 30 = 1200$ units (English). A force of 250 dynes acting 20 seconds would produce a change of momentum equal to $250 \times 20 = 5000$ units (C. G. S.).

QUESTIONS ON NEWTON'S SECOND LAW.

1. What acceleration would be produced by a force of 30 poundals acting on a mass of 80 pounds for 70 seconds ?

2. What acceleration would be produced by a force of 240 dynes acting on a mass of 3 kilogrammes for 3 minutes ?

3. What must be the intensity of a force that would give 90 pounds an acceleration of 1000 feet in 20 seconds ?

4. What must be the intensity of a force that would give a mass of 80 grammes an acceleration of 50 metres in 2 minutes ?

5. What must be the mass of a body to which a force of 60 poundals would give an acceleration of 500 feet in 30 seconds ?

6. What must be the mass of a body to which a force of 500 dynes would give an acceleration of 8 decimetres in 8 seconds ?

7. What momentum would be imparted to a body by a force of 70 poundals in 90 seconds ?

8 What momentum would be imparted to a body by a force of 350 dynes in 75 seconds ?

9. What force would be needed to change the momentum of a body 300 units (English) in 9 seconds ?

10. What force would be needed to change the momentum of a body 900 units (C. G. S.) in 60 seconds?

11. How long will it take a force of 120 poundals to impart a momentum of 700 units to a body ?

12. How long would it take a force of 600 dynes to impart a momentum of 19,000 units to a body ?

13. How long would it take a force of 20 poundals, acting in the opposite direction to the motion of the body, to stop a body having a momentum of 300 units ?

14. How long will it take a force of 80 dynes, acting in the opposite direction to the motion of the body, to stop a body having 1000 units of momentum ?

29. *Parallelogram of Motion.* — To find the path of a body *A* (Figure 5) acted on by two forces at the same time, draw *A B* to represent the path the body would have taken had it been acted on by the first force alone, and *A C* to represent the path it would have taken had it been acted on

Fig. 5.

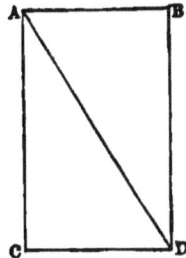

by the other force alone. Through B draw BD parallel to AC, and through C draw CD parallel to AB so as to complete the parallelogram $ABDC$. Draw the diagonal AD. This diagonal will represent the path taken by the body when acted upon by both forces together.

30. *Newton's Third Law of Motion.* — Newton's third law of motion is as follows: *Reaction is always equal and opposite to action; that is to say, the actions of two bodies upon each other are always equal and in opposite directions.*

This law simply states the fact that a force always acts upon two portions of matter (19), and that the stress, whether that of tension or pressure, is equal upon both portions. A stone raised from the earth attracts the earth just as much as the earth attracts the stone. Gravity really acts as a stress of tension between the two, and pulls them equally but in opposite directions. When the stone falls the earth moves up to meet it. When the two meet they have each the same momentum, but the earth, owing to its great mass, has only a very small velocity. When a cannon is fired, the igniting powder pushes back upon the cannon just as hard as it pushes forward on the ball. Were the cannon as free to move as the ball, it would start back, or *recoil*, with the same momentum that the ball starts forward with, but of course with a less velocity.

31. *Collision of Elastic Bodies.* — We have an illustration of action and reaction in the collision of elastic bodies. Place two

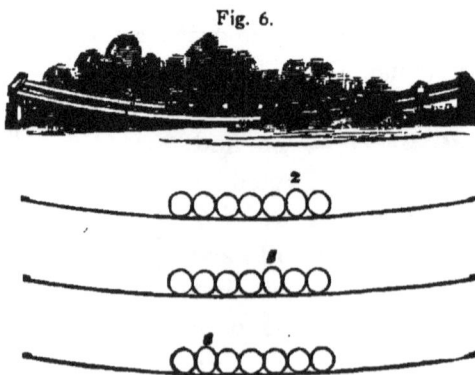

Fig. 6.

ivory balls of exactly the same size at the centre of the curved railway in Figure 6. Move one of the balls up to one end of the track, and let it roll back against the ball at rest. There will be a slight strain of compression when the balls strike, and this will develop a stress of elasticity between them which will act equally upon both and in opposite directions. This stress will stop the first ball and start the second off with the velocity the first had on striking it.

Place several ivory balls of the same size on the centre of the track, and allow the first ball to roll against the end of the line. All the balls will remain at rest except the last, which will be shot up the track. In this case the strain of compression and stress of elasticity have been propagated along the line from ball to ball. Each ball has been compressed a little in turn, and in recovering itself has pushed upon the ball behind it enough to stop it, and upon the one in front enough to flatten it a little. Each ball except the last was kept from moving forward by the reaction of the ball in front.

B. Work and Energy.

32. Work. — *Work* is said to be done when anything is *moved against resistance.* We may consider work either with reference to the force that moves the body, or with reference to the resistance overcome. When we think of the force as moving the body, we say that work is done by the *force* upon the body. When we think of the resistance as overcome by the body, we say that work is done by the *body* upon the resistance. When we think of the resistance as impeding the motion of the body, we say that work is done by the *resistance* upon the body.

These terms apply to different aspects of the same work. Thus, when we raise a weight, in winding up a clock, we may say that work is done by the *force* used upon the weight, or by the *weight* upon or against gravity, or by *gravity* upon the weight. The amount of work done is the same in whatever aspect we view it. When the clock weight runs down again, we may say that work is done by *gravity* upon the weight, or by the *weight* upon the resistance of the wheels, or by the *resistance of the wheels* upon the weight, according to the aspect in which we view the work. When a weight is allowed to fall freely to the earth, the work done is that of increasing the velocity of the body. In this case work is done by gravity *upon* the body, and *by* the body upon its inertia.

Work is done in every case in which *the velocity of a body is changed*, for the inertia of the body always resists this change. .

33. *Units of Work.* — The unit of work is *the work done in moving anything a unit of distance against a unit of resistance, or by a unit of force acting through a unit of distance.* A resistance is, of course, merely the opposing action of some force, and is measured in poundals or dynes. The *English unit of work* is the work done in *moving a mass against a poundal of resistance,* or by *the force of a poundal acting one foot.* It is called a *foot-poundal.* The *C. G. S. unit of work* is *the work done in moving a mass one centimetre against a dyne of resistance,* or by *the force of a dyne acting one centimetre.* It is called an *erg.* There are 421,393.8 ergs in a foot-poundal. These are *absolute* units.

The *gravitation unit of work* is *the work done in raising a unit of mass a unit in height.* The *English* gravitation unit is the work done in *raising a pound one foot high.* It is called a *foot-pound.* It varies with the force of gravity in different parts of the earth and at different elevations. At Greenwich there are 32.2 foot-poundals in a foot-pound.

34. *Energy.* — *Energy* is the *capacity for doing work.* It is measured in the same units as work, a *unit of energy* being the *capacity for doing a unit of work.* Thus, we may speak of so many foot-poundals, or of so many ergs, of energy.

The force that tends to stop a moving body acts upon it as a resistance, and every moving body has the power to overcome this resistance through a greater or less distance according to its momentum and velocity. Hence every moving body has a capacity for doing work, or *energy.* A body which is not in motion may have a capacity for doing work growing out of its condition with respect to some force. Thus, a raised weight has the ability to drive a clock, compressed steam the ability to drive a locomotive, and a coiled spring the ability to drive a watch.

35. *Position of Advantage.* — A body is said to have a *position of advantage* with respect to a force when it is *so situated that it is possible for that force to put it in motion.* A weight raised from the earth has a position of advantage with respect to gravity, since it is possible for gravity to put it in motion by pulling it to the earth again. For a similar reason molecules when separated from each other have positions of advantage with respect to cohesion ; and atoms when separated from each other, with respect to affinity. A strained body has a position of advantage with respect to elasticity.

36. *Kinetic Energy.* — The *energy of motion* is called *kinetic energy.*

The kinetic energy of a body is equal to *the product of the momentum of the body and ½ its velocity.* Now we may regard this work either as work done *by* the force acting as a resistance *upon* the body or *by* the body *upon* the resistance.

37. *Potential Energy.* — The *energy of position* is called *potential energy.* It is universally true that a body, in returning from a position of advantage to its original position, *does exactly the same amount of work that was done upon it* in putting it in its position of advantage. Thus, to raise a pound weight 12 feet high requires 12 foot-pounds of work. The same weight, in falling 12 feet, will do 12 foot-pounds of work. If it takes 300 ergs of work to coil a spring, the spring in uncoiling will do 300 ergs of work. Hence the potential energy of a body is equal to *the work required to put the body in its position of advantage.*

15. How many foot-poundals of energy has a mass of 2500 pounds with a velocity of 5000 feet a second?

16. How many ergs of energy has a mass of 8965 grammes with a velocity of 8000 centimetres a second?

17. How many foot-poundals of energy has a mass of 3 tons with a velocity of 500 feet a second?

18. How many ergs of energy has a mass of 9 kilogrammes with a velocity of 8 metres a second?

C. COMPOSITION AND RESOLUTION OF FORCES.

38. *Representation of Forces by Lines.* — A force may be completely represented by a line ; the *length* of the line representing the *intensity* of the force, the *direction* of the line the *direction* in which the force acts, and one *end* of the line the *point of application* of the force.

39. *Resultant and Component Forces.* — There is usually some one force that would have the same effect upon a body, in producing pressure or motion, as that of the several forces that may be acting together upon it. This force is called the *resultant* of these forces, and they are called its *components*.

40. *Composition and Resolution of Forces.* — The *combining of several forces into one resultant* is called the *composition of forces ;* and the *decomposition of one force into two or more components*, the *resolution of forces.*

In the composition and resolution of forces it is necessary to find the intensity, the direction, and the point of application of the resultant or components.

41. *The Parallelogram of Forces.* — Of the great variety of cases that may occur in the composition of forces, the most important is that in which two forces act upon a point in different directions. For example, let the two forces *A B* and *A C* (Figure 7) be acting upon the point *A* in the directions indicated by the arrows. Through

B draw the line $B D$ parallel to $A C$; and through C, the line $C E$ parallel to $A B$, so as to form the parallelogram $A B R C$. The diagonal $A R$ of this parallelogram will be the resultant of these two forces. This method is called the *parallelogram of forces*.

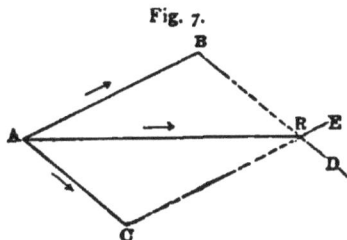

Fig. 7.

If a force $A G$ (Figure 8), having the intensity of the result-ant $A R$, but the opposite direction, were applied to A, it would balance this resultant, and, there-fore, its components $A B$ and $A C$.

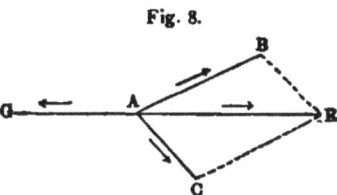

Fig. 8.

The fact that the resultant of forces may be balanced by an equal force applied to the same point in the opposite direction en-ables us to find the resultant of forces experimentally, and so to verify the above method. The apparatus for this experimental

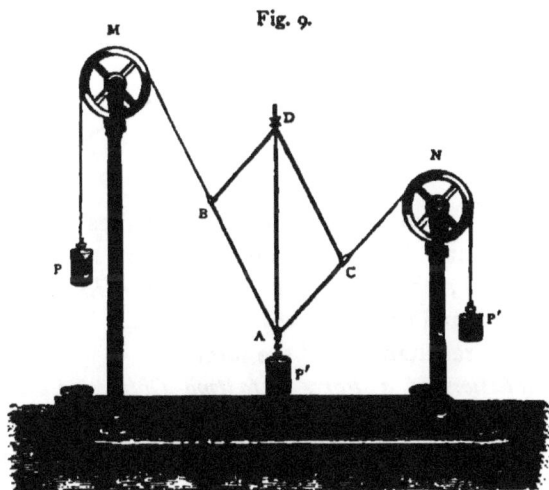

Fig. 9.

determination is shown in Figure 9. $A B D C$ is a parallelogram jointed at its four corners. Cords pass from the corners B and C over the pulleys M and N. Weights P and P' are attached to the ends of these cords. The number of ounces in the weight P is equal to the number of inches in the side $A B$; and the number of ounces in P', to the number of inches in $A C$. Hang from A a weight P'', less than the sum of P and P'. The parallelogram will take up a position of equilibrium such that the cords attached to B and C will be found to form prolongations of the sides $A B$ and $A C$, and the diagonal $A D$ will be vertical. The number of inches in the diagonal will be found to be equal to the number of ounces in the weight hung from A. The two forces P and P' which are acting on the point A are represented by the lines $A B$ and $A C$, and their resultant by the diagonal $A D$. This vertical resultant is balanced by the equal force P'' acting in the opposite direction.

42. *Composition of Several Forces acting in Different Directions upon a Point.* — When more than two forces, $A B$, $A C$, $A D$, and $A E$ (Figure 10) are acting in different directions upon a point A, their resultant may be found by the following method : —

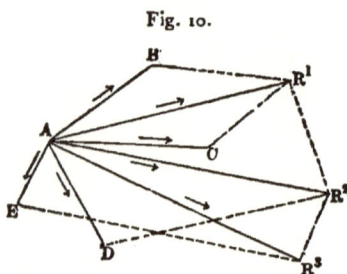

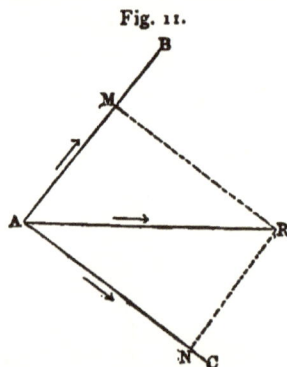

Fig. 10.

Fig. 11.

First find the resultant $A R^1$ of the two forces $A B$ and $A C$; then the resultant $A R^2$ of the first resultant $A R^1$ and of the third force $A D$; and, finally, the resultant $A R^3$ of the second resultant $A R^2$ and of the fourth force $A E$. This last resultant will be the resultant of all the forces.

43. *Resolution of a Force into two Oblique Forces.* — To resolve the force $A R$ (Figure 11) into two forces having the directions $A B$ and $A C$, draw $R M$ parallel to $A C$, and $R N$ parallel to $A B$. $A N$ and $A M$ will represent the forces required.

The resolution of forces may be illustrated by the case of a vessel sailing in any other direction than that of the wind.

Let the line $A\,B$ (Figure 12) represent the direction of the keel of the vessel; the line $C\,D$, the direction of the face of the sail; and the line $W\,E$, the direction and intensity of the wind. To find the intensity of the force which would be effective in driving the vessel forward, first resolve the force of the wind $W\,E$ into two components, one $D\,E$ tangent to the sail, and the other $F\,E$ perpendicular to the sail. This latter component will be the only part of the force of the wind that will have any effect upon the sail. This force must again be resolved into two components, one $G\,E$ perpendicular to the length of the vessel, and the other $H\,E$ in the direction of the vessel. This last component will be the only portion of the force of the wind that will be effective in moving the vessel forward.

Fig. 12.

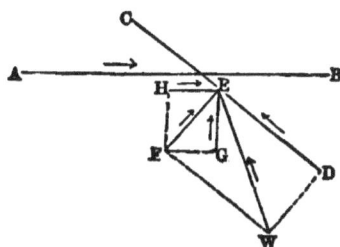

D. GRAVITY AND EQUILIBRIUM.

- **44. *Law of Gravity.*** — The law of gravity was discovered by Newton. It is as follows: *Every portion of matter attracts every other portion of matter with a force directly proportional to the product of the masses acted upon, and inversely proportional to the square of the distances between them.*

45. *Centre of Gravity.* — The direction of gravity at the surface of the earth is that of a plumb-line. Gravity acts upon every particle of which a body is composed, but the parallel forces acting upon the various particles may be resolved into one, and the point of application of this resultant is called the *centre of*

Fig. 13.

gravity of the body.   Thus, *G* is the centre of gravity in the stone in Figure 13.   The whole of the force of gravity act- ing upon a body may be considered as applied at the centre of gravity.   If a force equal to the resultant of the forces of gravity is applied to the centre of gravity in the opposite di- rection, the body will balance or be in equilibrium.

The centre of gravity may therefore be defined as *the point upon which the body will balance in every position.*

When a body is homogeneous throughout, the centre of grav- ity is at the centre of figure of the body.   When the body is not homogeneous throughout, the centre of gravity is away from the centre of figure towards the denser side of the body. The centre of gravity often lies entirely outside of the material of the body, as in the case of a ring or a hollow sphere.   When this is the case, the centre of gravity must be rigidly connected to the body in order to have the body balance on it.   A system of bodies may have a common centre of gravity lying outside of all the bodies.   The centre of gravity of two spheres will lie somewhere on a line between their centres of gravity.   If the spheres have the same mass, this point will lie just midway between their centres of gravity.   If one sphere has a greater mass than the other, the centre of gravity of the system will lie nearer the centre of gravity of the larger sphere.   If there is sufficient difference between their masses, their common centre of gravity may lie within the larger sphere.

46. *Experimental Method of finding the Centre of Grav-*

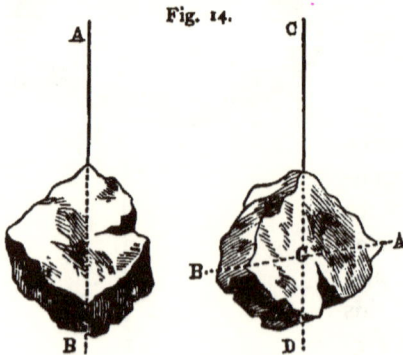

Fig. 14.

ity. — Since the result- ant of the forces of grav- ity acting upon a body, and the force which bal- ances it, must act along the same line in opposite directions, the centre of gravity of a body sus- pended so as to turn freely must be *in a ver-*

tical line under the point of suspension. Hence, if we suspend any body from two points and mark the vertical lines from each point of suspension (Figure 14), the centre of gravity must be where these verticals cross.

47. *Kinds of Equilibrium.* — When a body, on being tipped a little, tends to return to its old position, it is said to be in *stable* equilibrium ; when it tends to fall to a new position, in *unstable* equilibrium ; and when it rests equally well in every position, in *indifferent* equilibrium.

When a body is in stable equilibrium, its centre of gravity *rises* on tipping the body ; when it is in unstable equilibrium, its centre of gravity *falls* on tipping the body ; and when it is in indifferent equilibrium, its centre of gravity *neither rises nor falls* on tipping the body.

Fig. 15.   Fig. 16.

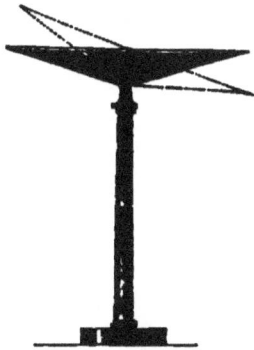   

48. *Equilibrium of a Body resting on a Fixed Point or Axis.* — A body resting on a point or axis can be in equilibrium only when the centre of gravity and the point or axis of support lie in the same vertical line. This can be the case only when the centre of gravity is *either directly above or below the point or axis of support.* In the former case the body is in *unstable* equilibrium. This case is shown in Figure 15. *O* is the axis of support, and *G* the centre of gravity. It will be seen that gravity will tend to

topple the body over as soon as it is tipped.   In the latter
case the body is in *stable* equilibrium.   This case is shown
in Figure 16.   As soon as the body
is tipped gravity tends to right it.

Fig. 17.

The toy called the *balancer* (Figure
17) is an illustration of stable equilibrium in a body resting on a point.
The balls at the ends of the wires at
each side of the figure bring the centre
of gravity of the whole below the toe
on which the figure is resting.

In a similar way a
cork may be balanced
on the point of a needle
by sticking two forks
into it, as shown in
Figure 18.

Fig. 18.

When the centre of
gravity is at the point
or axis of support, the body is in *indifferent*
equilibrium.

49. *Equilibrium of a Body resting on a Horizontal Plane
at One Point only.* — Such a body can be in equilibrium
only when its centre of gravity and the point where it

Fig. 19.

touches the plane are both in the same vertical.   Figure
19 represents two positions of equilibrium of an oval body
on a horizontal plane.   In the first case the body is in
unstable equilibrium, because its centre of gravity *will*

begin to fall as soon as it is tipped. In the second case the body is in *stable* equilibrium, because its centre of gravity is *in its lowest possible position.*

Fig. 20.

The toy called the *tumbler* (Figure 20) is an illustration of stable equilibrium of a body touching a horizontal plane at one point. The centre of gravity is so low down that the body cannot be tipped without raising this point.

50. *Equilibrium of a Body resting on a Horizontal Plane at Several Points.* — Such a body will be in stable equilibrium when the vertical line from its centre of gravity passes *within the polygon formed by joining the several points on which the body rests,* as in Figure 21. This polygon is called the *base* of the body. The lower the centre of gravity, and the greater the distance of its vertical from the nearest side of the base, the greater the stability of the equilibrium of the body, because the farther the body would have to be tipped, and the more its centre of gravity would have to be raised, to overturn it (Figure 22).

Fig. 21.

In Figure 22, in order to overturn either of the bodies *a b c d,* we must tip it so as to carry its centre of gravity *g* through the arc *g e* or raise it through the distance *h e;* and it is evident that *g e* and *h e* are greater in the case of the right-hand body.

For this reason a high load on a wagon is more likely to tip over than a low one. A leaning body, like the famous Leaning Tower at Pisa, may be in stable equilibrium, because the vertical from the centre of gravity falls within the base.

Fig. 22.

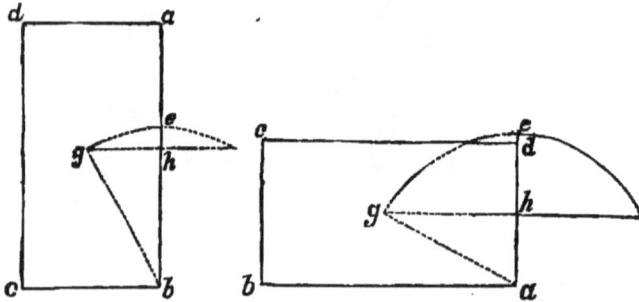

51. *Weight.* — Weight is *the downward pressure which gravity causes a body to exert.* While a body will have the same mass wherever it may be, its weight will vary with the force of gravity acting upon it. As this force is inversely proportional to the square of the distance (44), at twice the distance from the centre of the earth a body would have only one fourth the weight it has at the surface of the earth. On the sun, which has a much greater *mass* than the earth (44), the same body would have 28 times the weight it has on the earth. The English unit of weight is the *pound avoirdupois;* the French unit is the *gramme.*

The weight of a body is ascertained either by finding how much it will bend a spring, as in the spring balance, or by finding how many known weights at one end of a beam will counterpoise it when placed on the other end, as in the ordinary balance. By the last method the weight of the body would be found to be the same everywhere, for it is not the weight of the body which is found in this case, but its *mass.* This is found by comparing the weight of a body with that of a known mass. The weight of the mass to be weighed, and that of the mass used to counterpoise it, both change with the force of gravity.

52. *The Balance.* — The *balance* (Figure 23) consists of a rigid bar *A B*, called the *beam*, supported on an axis *O* at its centre. This axis is just above the centre of gravity of the beam, that the beam may be in stable equilibrium. When the beam is exactly horizontal, it is in equilibrium. Two scale pans are suspended from the ends of the beam at equal distances from the axis. The body to be

Fig. 23.

weighed is placed in one pan, and is counterpoised by known weights in the other.

53. *Specific Gravity.* — The *specific gravity* of a substance is *its weight compared with the weight of the same bulk of some standard substance.* The substance commonly taken as the standard for solids and liquids is *distilled water* at a temperature of 39° F. A cubic foot of such water weighs 62.425 lbs. avoirdupois. The weight of a gallon of water is 10 lbs. The weight of a cubic centimetre of water is a gramme, and the weight of a litre of water is a kilogramme.

In the following table we give the specific gravities of some liquids and solids.

Liquids, at Temperature of Freezing Water.

Water, sea, ... 1.026	Oil, linseed 940
Alcohol, pure 791	" olive 915
" proof spirit . .916	" whale 923
Ether 716	" turpentine 870
Mercury 13.596	Blood, human 1.055
Naphtha 848	Milk, of cow 1.03

Solids.

Copper 8.95	Ice 92
Gold 19.26	Basalt . . . . : . 3.00
Iron 7.79	Clay 1.92
Iridium 22.4	Glass, crown . . . 2.5
Lead 11.4	" flint 3.0
Platinum 21.5	Quartz (rock crystal) . 2.65
Silver 10.5	Fir, spruce . .. . .48 to .7
Tin 7.3	Oak, European . .69 to .99
Zinc 6.9	Lignum-vitæ . . 65 to 1.33

The weight of a cubic foot of any substance is equal to 62.425 lbs. avoirdupois multiplied by its specific gravity.

The weight of a cubic centimetre of any substance, in grammes, is equal to its specific gravity.

The weight of a litre (or cubic decimetre) of any substance, in kilogrammes, is equal to its specific gravity.

The weight of a gallon of any liquid, in lbs. avoirdupois, is equal to its specific gravity multiplied by 10.

QUESTIONS ON THE ABOVE TABLE.

19. What is the weight of a cubic foot of mercury?
20. What is the weight of a gallon of milk?
21. How many gallons in 50 lbs. of pure alcohol?
22. What is the weight of 15 litres of ether?
23. How many litres in 8 kilogrammes of olive oil?
24. What is the weight of a cubic yard of clay?
25. What is the weight of a cubic foot of flint glass?
26. What is the weight of a cubic inch of silver?
27. How many cubic feet in a ton of ice?
28. How many cubic inches in a pound of quartz?
29. What is the weight of a cubic decimetre of silver?
30. What is the weight of a cubic metre of lead?

E. FALLING BODIES.

54. *All Bodies fall at the Same Rate in a Vacuum.* — That light and heavy bodies fall at the same rate in a vacuum may be shown with the guinea and feather tube

(Figure 24). The tube contains a bit of metal and a feather. Exhaust the air from the tube, and invert the tube. The metal and the feather will be seen to fall through the tube at the same rate.

Fig. 24.

The reason that light and heavy bodies fall in a vacuum at the same rate is that the force of gravity acting upon a body varies directly as the mass of the body. The force of gravity on a mass of a pound is about 32 poundals; on a mass of two pounds, 64 poundals; on a mass of half a pound, 16 poundals; on a mass of one ounce, 2 poundals; etc. The force of gravity on a mass of one gramme is about 981 dynes; on a mass of a decagramme, 9810 dynes; on a mass of a decigramme, 98.1 dynes; etc. Since the intensity of the gravity acting upon a body increases just as rapidly as the mass of the body, gravity, if left to itself, would cause all bodies to fall at the same rate; for if the mass of one body is twice or thrice as great as that of another, gravity will act upon it with twice or thrice the intensity.

55. *Bodies fall with Unequal Veloci-ties in the Air.* — A bullet will fall through the air much faster than a feather. The air *offers resistance to every body falling through it.* The denser a body and the less its surface, the less its motion is re-tarded by the air. Gold-leaf falls slowly in the air, while the same gold in the form of a solid sphere would fall almost as rapidly in the air as in a vacuum.

The resistance of the air increases with the velocity, and after a while it becomes equal to the attraction of gravity upon a body. When this is the case, the body will gain no more velocity, but keep falling at a uniform rate. Were a body shot downward with a velocity greater than this, it would be retarded by the resistance of the air, which would then be greater than the pull of gravity, until its velocity were reduced to that at which the resistance of the air would be just equal to the pull of gravity.

56. *Acceleration produced by Gravity.* — When bodies are falling near the earth, *gravity increases their velocity at the uniform rate of about* 32.2 *feet a second*, in a vacuum. This acceleration per second produced by gravity is usually represented by *g*, and is called the *intensity* of gravity. It is equal to about 981 centimetres, or 9.81 metres. When a body is rising, gravity *retards* its velocity at the rate of 32.2 feet, or 9.81 metres a second. Were a body thrown up in a vacuum, it would be just as many seconds in falling as it is in rising, and it would reach the point it started from with the velocity it had on starting. It gains just as much velocity in falling as it lost in rising.

The *velocity* acquired by a body falling from a state of rest will be equal to *the product of the intensity of gravity and the number of seconds the body has been falling.* If we represent the velocity acquired by *v*, and the number of seconds the body has been falling by *t*, the formula for the velocity of a body falling from a state of rest will be $v = g t$.

If a body were falling from a state of rest, the number of feet of velocity it would acquire in 20 seconds would be 32.2 × 20 = 644 ; and the number of metres of velocity it would acquire would be 9.81 × 20 = 196.2.

The *distance* passed over by a moving body is always equal to *the product of its mean velocity and the time.* Since falling bodies gain velocity at a uniform rate, the *mean velocity* of a body falling from a state of rest will be *one half the velocity it has acquired.* As the velocity is $= g t$, the mean velocity will be ½ *g t*. If we represent the space passed over by *s*, we shall have

$$s = \tfrac{1}{2} g t \times t = \tfrac{1}{2} g t^2.$$

The distance passed over by a body falling 4 seconds from

a state of rest would be equal to 16.ᵻ × 16 = 257.6 feet, or to 4.9 × 16 = 78.4 metres.

From the formula

$$v = g t$$

we have

$$t = \frac{v}{g}.$$

Substituting this value of t in the formula

$$s = \tfrac{1}{2} g t^2,$$

we have

$$s = \frac{v^2}{2 g}.$$

QUESTIONS ON FALLING BODIES.

31. How long would it take a body falling from a state of rest to acquire a velocity of 193.2 feet ?

$$t = \frac{v}{g} = \frac{193.2}{32.2} = 6 \text{ seconds.}$$

32. How long would it take a body falling from a state of rest to acquire a velocity of 39 24 metres a second ?

$$t = \frac{v}{g} = \frac{39.24}{9.81} = 4 \text{ seconds.}$$

33. How far must a body fall from a state of rest to acquire a velocity of 1500 feet a second ?

$$s = \frac{v^2}{2 g} = \frac{2250000}{64.4} = 34938 \text{ feet.}$$

34. How far must a body fall from a state of rest to acquire a velocity of 800 metres a second ?

$$s = \frac{v^2}{2g} = \frac{640000}{19.62} = 32619.7 \text{ metres.}$$

35. How many feet of velocity would a body acquire in falling 25 seconds from a state of rest ?

36. How many metres of velocity would a body acquire in falling 42 seconds from a state of rest ?

37. How long would a body have to fall from a state of rest to acquire a velocity of 986 feet ?

38. How long would a body have to fall from a state of rest to acquire a velocity of 25,000 centimetres a second ?

39. How many feet would a body fall from a state of rest in 32 seconds ?

40. How many metres would a body fall from a state of rest in 45 seconds?

41. How far would a body have to fall from a state of rest to acquire a velocity of 1200 feet a second ?

42. How far would a body have to fall from a state of rest to acquire a velocity of 300 metres a second ?

57. *The Height to which a Body can rise.* — A body moving upward will continue to rise till all of its velocity is exhausted. A rising body loses velocity just as fast as a falling body gains it. Hence the height to which a body can rise with a given velocity is just *equal to the height from which it must fall to gain that velocity.* The height to which a body can rise will therefore be represented by the formula

$$s = \frac{v^2}{2g}.$$

In this case s is the distance a body can rise, and v the velocity with which it starts. The height to which a body can rise *increases as the square of the velocity with which it starts.*

58. *Transformation of Energy in the Case of a Body projected upward.* — When a body is projected upward, its energy on leaving the surface of the earth is entirely kinetic. As it rises, it moves slower and slower, and so loses kinetic energy, but as it is separated farther and farther from the earth, it gains potential energy. At the highest point the body reaches, its energy is entirely potential. As it falls again, it moves faster and faster, and so gains kinetic energy, but as it comes nearer and nearer the earth, it loses potential energy. While the body is rising its kinetic energy is gradually transformed into potential energy ; and when it falls again, its potential energy is changed back again into kinetic energy. The energy possessed by the body is precisely the same at every point in its path. When the body strikes the earth, its energy is apparently destroyed ; but when we come to the subject of Heat, we shall see that this is not really the case.

59. *The Path of a Body projected horizontally or obliquely.* — When a body is projected horizontally or obliquely, gravity draws it towards the earth faster and faster the longer it acts upon it, and so *causes it to describe a curved path.* The curve in this case would be a *parabola* were it not for the resistance of the air.

The curved line in Figure 25 shows approximately the path of a cannon-ball through the air, when fired in the direction of *A B.* The line *A C* represents the *range* of the ball, or the greatest horizontal distance it is thrown. Were it not for the resistance of the air, the range would be greatest when the cannon was pointed 45° above the horizon.

Fig. 25.

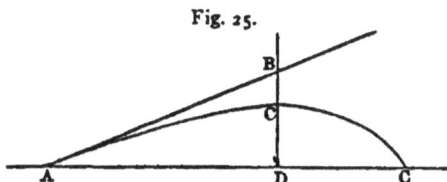

60. *Intensity of Gravity.* — The intensity of gravity *varies as we pass from the equator to the poles.* At the equator its intensity is sufficient to give a mass in a vacuum an acceleration of 32.088 feet per second, while at the poles it is sufficient to give a mass in a vacuum an acceleration of 32.253 feet per second. The value of *g* in centimetres varies from 978.10 at the equator to 983.11 at the poles. The intensity of gravity also *varies with the height* (44). At twice the distance from the centre of the earth, the intensity of gravity is only one fourth as great ; at three times the distance, one ninth as great ; and so on.

Since a *poundal* is a force that will give to a mass of a pound an acceleration of a foot in a second, and since gravity will give a mass of a pound an acceleration of 32.2 feet a second, it follows that there are about 32.2 poundals in a pound at Greenwich as has already been stated (33). A poundal is about half an ounce. The number of poundals in a pound at any place is equal to the value of *g* in feet at that place.

Since gravity will give a mass of a gramme an acceleration of 981 centimetres, it follows that there are 981 dynes in a gramme of force at Greenwich. The number of dynes in a gramme at any place is equal to the value of *g* in centimetres at that place.

The value of g at any place is ascertained by means of a pendulum.

F. The Pendulum.

61. *The Pendulum.* — Any body free to turn on a horizontal axis which does not pass through its centre of gravity can be in stable equilibrium only when its centre

Fig. 26.

of gravity is below the axis of support and in the same vertical plane with it. When pulled aside from this position of equilibrium and released, the body *will vibrate to and fro across its position of stable equilibrium*, until friction and the resistance of the air bring it to rest. A body suspended in this way, no matter what its shape, is called a *pendulum*. The usual form of the pendulum is that shown in Figure 26. It consists of a rod which can turn on an axis O at its upper end, and which carries a heavy piece of metal M, called the *ball*, at its lower end. The ball can be raised or lowered by means of the screw V.

The *arc described* by the pendulum is called the *amplitude* of the vibration, and the *time it takes to describe it* is called the *time* of vibration.

62. *The Laws of the Pendulum.* — It has been found, by mathematical investigation, that *for small vibrations the time of vibration is independent of the amplitude ;* also, that *the time of vibration increases as the square root of the length of the pendulum*, and *decreases as the square root of the intensity of gravity increases.* In other words, when the amplitude does not exceed $3°$ or $4°$, the same pendulum will vibrate at the same rate, no matter what may be the amplitude of vibration ; but if the pendulum is made four, nine, or sixteen times as long, it will vibrate one half, one third, or one fourth as fast ; while, if a pendulum were kept of the same length, and the

intensity of gravity were to become four, nine, or sixteen times as great, the pendulum would vibrate two, three, or four times as fast.

63. *Simple and Compound Pendulums.* — The *simple pendulum* is an ideal one, whose ball is a single heavy particle suspended by a line without weight. Every pendulum actually used is a *compound* one, consisting of a heavy weight hung from a fixed point by means of a rod of wood or metal. Each particle of such a pendulum may be regarded as a simple pendulum ; but as these particles are at different distances from the point of suspension, they tend to vibrate at different rates. The particles near the point of suspension are retarded by the tendency of the particles below them to vibrate at a slower rate, while the particles near the lower end of the pendulum are accelerated by the tendency of the particles above them to vibrate more rapidly. At some point between these there must be *a particle whose vibration is neither retarded nor accelerated.* As this particle is vibrating at its normal rate, *its distance from the point of suspension* must be *the length of a simple pendulum that would vibrate at the rate of the compound pendulum.* The point where this particle is situated is called the *centre of vibration ;* and its distance from the point of suspension, the *virtual length* of the pendulum.

If a pendulum is inverted and suspended by its centre of vibration, *the former point of suspension becomes its new centre of vibration.* This remarkable property of a compound pendulum enables us readily to find the centre of vibration. We have only to reverse the pendulum, and find, by trial, the point at which it must be suspended to vibrate at the same rate as before. A pendulum constructed for this purpose is called a *reversible pendulum.*

64. *Use of the Pendulum for measuring Time.* — The most important use of the pendulum is for *measuring time.* A common *clock* is an instrument for *keeping a pendulum in vibration, and recording its beats.*

The essential parts of the arrangement by which this is accomplished are shown in Figure 27. The *scape-wheel R* is turned by a weight or spring, and its motion is regulated by means of the *escapement m n*. This turns on the axis *o*, and is connected

Fig. 27.

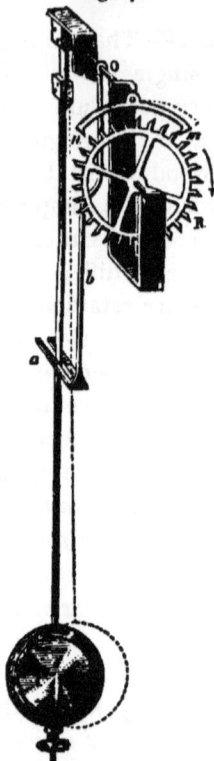

with the pendulum rod by means of the forked arm *a b*. When the pendulum is at rest, the hooks *n* and *m* of the escapement catch the teeth of the scape-wheel, and keep it from turning. As the pendulum vibrates, the hooks of the escapement alternately release and catch the teeth of the scape-wheel, and so compel it to turn slowly, and at a uniform rate. The hooks of the escapement are of such shape that each tooth of the scape-wheel, as it slips off the hook, gives the escapement a little push so as to keep up the vibration of the pendulum.

Each tooth of the scape-wheel is caught twice during the revolution of the wheel, once by each hook of the escapement. Hence, if the scape-wheel has thirty teeth, it will make one revolution for every sixty beats of the pendulum. The axis of the scape-wheel carries the *second-hand* of the clock, which registers the beats of the pendulum up to sixty. The scape-wheel is connected with another which turns $\frac{1}{60}$ as fast. The axis of this wheel carries the *minute-hand*, which registers the revolution of the second-hand up to sixty. This second wheel is connected with a third which turns $\frac{1}{12}$ as fast as itself. The axis of this last wheel carries the *hour-hand*, which registers the revolution of the minute-hand up to twelve, or half a day.

65. *Transformations of Energy in the Vibrations of the Pendulum.* — When the pendulum reaches its farthest point to the right or left, its energy is entirely *potential;* and when its ball is at its lowest point, its energy is entirely *kinetic.* As the ball rises, its kinetic energy is transformed into potential energy, and as it falls again, its potential energy is transformed into kinetic energy.

The energy consumed in overcoming the friction of the axis

of the pendulum and of the wheels of the clock and the resistance of the air is supplied by the falling weight or uncoiling spring ; and when the store of energy in the weight or spring is consumed, it must be renewed by again raising the weight or coiling the spring in winding up the clock. This new supply of energy is drawn from the hand and arm of the person who winds the clock.

G. Machines.

66. *Simple Machines.* — A *machine* is an instrument *by which a force is applied to do work.* Every machine, however complicated, is made up of a very few elements, called *simple machines,* or *mechanical powers.* These are the *lever,* the *wheel and axle,* the *pulley,* the *inclined plane,* the *wedge,* and the *screw.*

The *force applied* to work the machine is called the *power ;* and the *resistance overcome* by the machine, the *work.*

A perfect machine would be one which offered no friction or other resistance of its own. Such a machine has only an ideal existence. In every machine in actual use the work done is partly *useful* in overcoming the resistance we desire to overcome, and partly *useless* in overcoming the resistance of the machine itself. In the theory of machines the resistance of the machine itself is left out of view. The magnitude of the resistance to be overcome is represented by a rising weight, and the magnitude of the power is usually represented by a falling weight. The *resistance* is often called the *weight.*

67. *The General Law of Machines.* — The work done *by* the power *upon* a machine, and the work done *by* a machine *upon* the resistance, are simply different aspects of the same work (32), and hence they are equal in amount. Now the work done by a *falling* weight is equal to *the product of the weight and the distance it falls,* and the work done in *raising* a weight is *the product of the weight and the distance it is raised.* If, then, we represent the work done by the power upon the machine by a falling weight, and the work done

by the machine upon the resistance by a rising weight, we arrive at the following general law of machines : *The power multiplied by the distance through which it moves is always equal to the weight multiplied by the distance through which it moves.*   This law is often stated thus : *In any machine, the power and weight will be in equilibrium when they are in the inverse ratio of their velocities.*

The following facts result from the general law of machines just stated : —

(1.) The faster the power moves, compared with the weight, the greater the weight it will balance.

(2.) When the power moves faster than the weight, it will balance a weight greater than itself ; and when it moves slower than the weight, it will balance a weight less than itself ; and when it moves just as fast as the weight, it will balance a weight equal to itself.

(3.) The power will balance a weight just as many times itself as its velocity is times that of the weight.

68.   *Gain and Loss of Power in a Machine.* — When, in any machine, the power balances a weight greater than itself, there is said to be a *gain of power*, or *mechanical advantage;* and when the power balances a weight less than itself, a *loss of power*, or *mechanical disadvantage.*

When there is a *gain of power* there is always a corresponding *loss of speed*, and when there is a *loss of power* there is a corresponding *gain of speed.*

A machine might be described as an instrument by which we change the *point* at which the power acts, the *direction* in which it acts, or the *rate* at which it acts.   The last change is the most important one effected by a machine.   When the machine causes the power to act upon the resistance at a slower rate than it would were it applied directly to it, there is a gain of power ; and when it causes it to act upon it at a quicker rate, there is a loss of power.   When the machine does not change the rate, there is neither gain nor loss of power.

QUESTIONS ON THE GENERAL LAW OF MACHINES.

43. In a machine, the power moves 25 inches while the weight is moving 35 inches. What weight would be balanced by 63 pounds of power?

If we denote the power by P, the weight by W, the velocity of the power by VP, the velocity of the weight by VW, and the distances passed over by the power and weight, respectively, by DP and DW, then we shall have, in the above example,

$$VP = \tfrac{5}{7} \, VW$$
$$P = \tfrac{7}{5} \, W$$
$$63 = \tfrac{7}{5} \, W$$
$$63 \div \tfrac{7}{5} = 45$$
$$W = 45 \text{ pounds.}$$

44. In a machine, a power of 27 pounds balances a weight of 45 pounds. How far does the power move while the weight moves 60 inches?

$$P = \tfrac{3}{5} \, W$$
$$VP = \tfrac{5}{3} \, VW$$
$$DP = \tfrac{5}{3} \times 60 = 100 \text{ inches.}$$

45. In a machine, the power moves 56 inches while the weight moves 21 inches. What power will balance a weight of 600 pounds?

46. In a machine, the power moves 35 inches while the weight is moving 63 inches. What weight will be balanced by 250 pounds of power?

47. In a machine, the power moves 15 centimetres while the weight moves 40 centimetres. What power will balance a weight of 90 grammes?

48. In a machine, the power moves 24 centimetres while the weight is moving 56 centimetres. What weight would be balanced by 130 grammes of power?

49. In a machine, a power of 28 pounds balances a weight of 49 pounds. How far will the power move while the weight moves 20 inches?

50. In a machine, a power of 40 pounds balances a weight of 32 pounds. How far will the weight move while the power is moving 30 inches?

51. In a machine, a power of 50 grammes balances a weight

of 80 grammes.   How far will the power move while the weight
is moving 15 centimetres ?

52. In a machine, a power of 81 grammes balances a weight
of 63 grammes.   How far will the weight move while the power
is moving 25 centimetres ?

69. *The Lever.* — The *lever* is *a rigid bar, capable of turn-*
ing upon a fixed point or axis. The point on which the lever

Fig. 28.

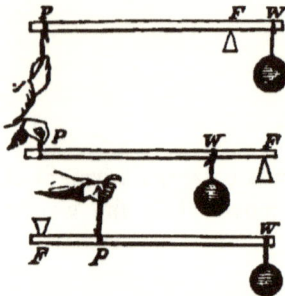

turns is called the *fulcrum.*

Different forms of the lever are
shown in Figure 28. *F* is the
fulcrum, *W* the weight, and *P* the
power.

When the *fulcrum* is between
the power and weight, the lever is
said to be of the *first order ;* when
the *weight* is between the fulcrum
and power, the lever is said to be
of the *second order ;* and when the *power* is between the
fulcrum and weight, the lever is said to be of the *third*
order.

Fig. 29.   Fig. 30.

A bar used for raising a weight is a lever.   When it is used
as shown in Figure 29, it is a lever of the first order.   When it

Fig. 31.

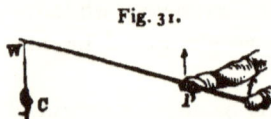

is used as shown in Figure 30, it is a
lever of the second order.   A fishing-
rod (Figure 31) is a lever of the third
order.

The *arms* of a lever are the *distances from the fulcrum to*
the points where the power and weight are applied, in case the
lever is straight ; or the distances from the fulcrum to the
lines which show the direction of the power and weight, in
case the lever is bent.

In Figure 28, FP is in each case the *power arm*, and FW the *weight arm*. In Figure 32 the dotted lines, which are supposed to be drawn from the fulcrum perpendicularly to the directions in which the weight and power act, are the arms of the bent lever, $abfc$.

Fig. 32.

70. *The Special Law of the Lever.* — The special law of the lever is that *the velocities of the power and weight are in the direct ratio of the lengths of the arms to which they are applied;* that is, if one arm of the lever is three times as long or one third as long as the other, the power or weight applied to this arm will move three times as fast or one third as fast as the one applied to the other arm.

There will be a *gain of power* in the lever whenever *the power arm is the longer;* for the power will then move the faster, and will balance a weight greater than itself. There will be a *loss of power* when *the power arm is the shorter;* for the power will then move the slower, and will balance a weight less than itself.

In a lever of the *second* order there will always be *a gain of power*, and in a lever of the *third* order *a loss of power*. In the lever of the *first* order there will be *a gain or loss of power, or neither*, according as the fulcrum is nearer the weight, or nearer the power, or midway between the two.

71. *The Compound Lever.* — Sometimes two or more simple levers are combined, as shown in Figure 33. If P is five times as far from the fulcrum f as A is, the point P will then move five times as fast as the point A, and a pull of one pound on P will exert a pull of five pounds on A. If B is five times as far from the fulcrum F as W is, the five pounds of pull on B will exert twenty-five pounds of pull at W. In this case one pound of pull exerted at P will balance twenty-five pounds at W; but it would be found on

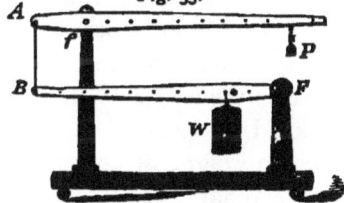

Fig. 33.

trial that by pulling *P* down one inch, *W* would be raised only one twenty-fifth of an inch.

Such a *combination of levers* is called a *compound lever*.

QUESTIONS ON THE LEVER.

53. In a lever, the power arm is 18 inches and the weight arm is 42 inches. What weight would be balanced by 60 pounds of power?

Denote the power arm by *P A*, and the weight arm by *W A*.

$$P A = \tfrac{3}{7} W A$$
$$V P = \tfrac{3}{7} V W$$
$$P = \tfrac{3}{7} W$$
$$60 = \tfrac{3}{7} W$$
$$60 \div \tfrac{3}{7} = 25\tfrac{5}{7}$$
$$W = 25\tfrac{5}{7} \text{ pounds.}$$

54. In a lever, the power arm is 36 inches, and the weight arm 27 inches. What power will balance a weight of 75 pounds?

55. In a lever, the power arm is 14 decimetres long, and the weight arm 21 decimetres. What weight would be balanced by 70 grammes of power?

56. In a lever, the power arm is 49 decimetres long, and the weight arm 28 decimetres. What power would balance a weight of 17 kilogrammes?

57. In a lever, a power of 30 pounds balances a weight of 50 pounds, and the power arm is 80 inches long. What is the length of the weight arm?

58. In a lever, a power of 70 pounds balances a weight of 20 pounds, and the weight arm is 30 inches long. What is the length of the power arm?

59. In a lever, a power of 150 grammes balances a weight of 250 grammes, and the power arm is 18 decimetres in length. What is the length of the weight arm?

60. In a lever of the first order, a power of 30 pounds balances a weight of 40 pounds, and the power arm is 27 inches long. What is the length of the lever?

61. In a lever of the first order, a power of 55 grammes balances a weight of 35 grammes, and the weight arm is 13 decimetres long. What is the length of the lever?

62. In a lever of the second order, the length of the lever is 65 decimetres, and a power of 24 grammes will balance a weight of 64 grammes. What is the length of the weight arm?

63. In a lever of the third order, the length of the lever is 28 decimetres, and the length of the power arm is 12 decimetres. What power will balance 18 grammes of weight?

72. *The Pulley.*—The *pulley* is a small *grooved wheel turning freely in a frame* called the *block*. It is a machine in which power is applied to do work by means of a *cord* instead of a bar, as in the case of the lever. The wheel of the pulley serves simply to diminish friction at the points over which the cord is drawn.

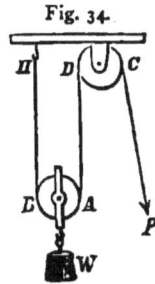

In Figure 34 the block of the pulley *D C* is fastened to the beam above, so as to be stationary, while the block of

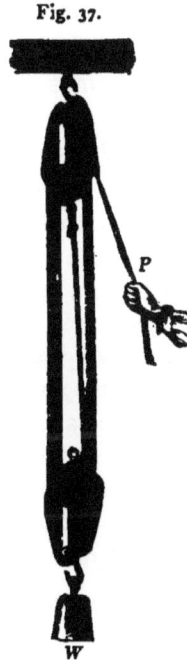

the pulley *A B* is free to move up and down. The former is called a *fixed* pulley; and the latter, a *movable* pulley. A *fixed* pulley serves simply to *change the direction in which the power acts.*

73. *Systems of Pulleys with one Cord.*—

In Figures 35, 36, and 37, are shown systems of pulleys
with a single cord, that is, in which one cord passes over
all the pulleys.   The power is applied to the end of the
rope, and the weight is attached to the movable block.
In the first case, on raising the movable block one inch,
three inches of rope will be released, since the rope comes
three times to that block.   In this case the power will
move three times as fast as the weight.   In the second
case, on raising the movable block one inch, four inches
of rope will be released, since the rope comes four times
to this block.   In this case the power will move four times
as fast as the weight.   In the third case the power will
move six times as fast as the weight.

The special law of a system of pulleys with a single
rope is that *the velocities of the power and weight are in the
inverse ratio of the number of times the cord comes to each.*
As the cord always comes once to the power, the power
will balance a weight as many times itself as the cord
comes times to the block bearing the weight.

QUESTIONS ON PULLEYS WITH SINGLE ROPE.

64.   In a system of pulleys with a single rope, the rope comes
13 times to the block bearing the weight.   What weight would
be balanced by 75 pounds of power?

65.   In a system of pulleys with a single cord, the cord comes
9 times to the block bearing the weight.   What power would
balance 19 grammes of weight?

66.   In a system of pulleys with a single rope, a power of 13
pounds balances a weight of 91 pounds.   How many times does
the rope come to the block bearing the weight?

67.   In a system of pulleys with a single rope, a power of 72
grammes balances a weight of 792 grammes.   How many times
does the rope come to the block bearing the weight?

74. *Systems of Pulleys with more than one Rope.* — The law
of the pulley is sometimes stated as follows : *A stretched rope
must have the same tension throughout its whole length.*

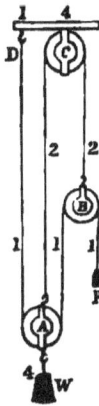

Figure 38 represents a system of pulleys in which two ropes are used. Here a weight of four pounds is balanced by a power Fig. 38. of one pound. The parts of the rope A D and A B must each have a tension equal to the power. The rope A C B balances the two tensions, B P and B A, and must therefore have a tension of twice the power. The three tensions supporting the pulley A amount therefore to four times the power.

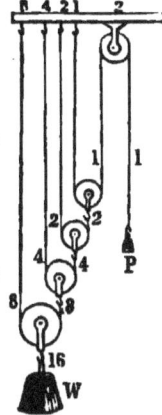

In the system shown in Figure 39 four ropes are used. The tensions of the several ropes will be readily understood from the numbers. It will be seen that in this case the power is doubled by each movable pulley which is added; but, as in all the systems we have examined, what is gained in power is lost in speed.

Fig. 39.

75. *Wheel and Axle.* — The *wheel and axle* (Figure 40) consists of a wheel, or drum, *a*, mounted on an axle *b*. The power and weight are applied to ropes which pass, one over the wheel and the other over the axle, in opposite directions, so that one unwinds as the other winds up. The power and weight are really applied to the wheel and axle at the point where the rope touches each, that is, at the end of the radius of each. The one applied to the wheel moves the faster, and just as many times faster as the circumference or the radius of the wheel is times the circumference or the radius of the axle.

Fig. 40.

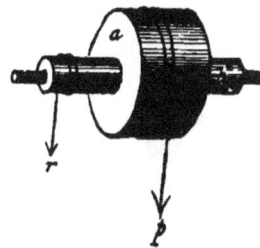

The special law of the wheel and axle is that *the velocities of the power and weight are in the direct ratio of the radii to which' they are applied.* When the power is applied to the wheel, there is a gain of power; and when it is applied to the axle, there is a loss of power.

The chief use of the wheel and axle in machinery is in transmitting rotary motion from one piece to another, with or without a change of velocity. For an increase of velocity, a large wheel must act upon a small one; and for a diminution of velocity, a small wheel must act upon a large one. When there is to be no change of velocity, the wheels must both be of the same size.

Fig. 41.

76. *Cog-Wheels.* — There are various ways in which the axle of one wheel is made to act on the circumference of another. Sometimes the one turns the other by rubbing against it, or by *friction*. The most common way, however, is by means of *teeth* or *cogs* raised on the surfaces of the wheels and axles. The cogs on the wheel are usually called *teeth*, while those on the axle are called *leaves*, and the part of the axle from which they project is called the *pinion*.

77. *The Gain of Power by Wheel-Work.* — In the train of wheels in Figure 41, if the circumference of the wheel *a* is 36 inches, and that of the pinion *b* is 9 inches, or one fourth as great, a power of one pound at *P* will exert a force of four pounds on *b*. If the circumference of the wheel *e* is 30 inches, and that of the pinion *C* 10 inches, the four pounds acting on the former will exert a force of twelve pounds on the latter. If the circumference of the wheel *f* is 40 inches, and that of the axle *d* 8 inches, the twelve pounds acting on *f* will exert a force of sixty pounds on *d*. One pound at *P* will then balance sixty pounds at *W*.

But in this case, as in all others, what is gained in power is lost in speed; since the one pound at *P* must move through sixty inches in order to raise the sixty pounds at *W* one inch.

78. *Belted Wheels.* — Another way in which wheels and axles may be made to act upon one another is by means of a *belt*, or band, passing over them both. They may thus be at any distance apart, and may turn either the same way or contrary ways,

Fig. 42.

Fig. 43.

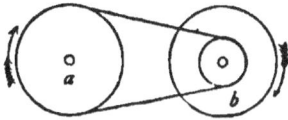
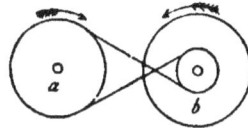

according as the belt does or does not *cross* between them (Figures 42 and 43). A cog-wheel and its pinion must, of course, always turn in contrary directions.

79. *The Windlass and Capstan.* — The *windlass* is a horizontal barrel turned by means of a crank or spokes (Figure 44). It may be regarded as a modification of the wheel and axle, the crank taking the place of the wheel. The *capstan* is an upright drum turned by means of levers, which may be removed at pleasure.

Fig. 44.

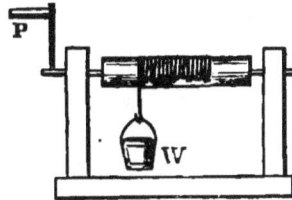

QUESTIONS ON THE WHEEL AND AXLE.

68. The radius of a wheel is 40 inches, and that of its axle 15 inches. What weight on the axle would be balanced by 50 pounds of power on the wheel?

Denote the radius of the wheel by $R\,W$, and that of the axle by $R\,A$.

$$R\,W = \tfrac{8}{3}\,R\,A$$
$$V\,P = \tfrac{8}{3}\,V\,W$$
$$P = \tfrac{3}{8}\,W$$
$$50 = \tfrac{3}{8}\,W$$
$$50 \div \tfrac{3}{8} = 133\tfrac{1}{3}$$
$$W = 133\tfrac{1}{3} \text{ pounds.}$$

69. The radius of a wheel is 18 decimetres, and that of its axle 12 decimetres. What weight on the wheel would be balanced by 32 grammes of power on the axle?

4

70. In a wheel and axle, a power of 63 pounds on the axle balances a weight of 35 pounds on the wheel. The radius of the wheel is 16 decimetres. What is the radius of the axle?

71. A power of 21 pounds on the wheel balances a weight of 77 pounds on the axle. The radius of the axle is 5 inches. What is the radius of the wheel?

80. *The Inclined Plane.* — An *inclined plane* is simply an *inclined surface.* It is easier to roll a body up an inclined surface than to raise the body vertically to the same height. The reason is obvious. The body must be raised against the action of gravity ; and by rolling the body up the inclined surface, the power is compelled to act through a distance equal to the length of the surface in raising the weight the height of it.

The special law of the inclined plane is that *the velocities of the power and weight are in the ratio of the length of the plane to its height.* Since the power and weight are in the inverse ratio of their velocities, it follows that *the power will be to the weight as the height of the plane is to its length.*

The law of the inclined plane may be demonstrated by means of the apparatus represented in Figure 45. *R S* is a smooth piece of hard wood hinged at *R;* by means of a screw it can be clamped at any angle *x* against the curved support ; *a* is a metal cylinder, to the axis of which is attached a string passing over a pulley to a scale-pan *P*.

Fig. 45.

It is thus easy to ascertain by direct experiments what weight must be placed in the pan *P* in order to balance a roller of any given weight.

The line *R S* represents the *length,* *S T* the *height,* and *R T* the *base* of the inclined plane.

81. *The Wedge.* — Instead of lifting a weight by moving

it along an inclined plane, we may do the same thing by *pushing the inclined plane under the weight.* When used in this way the inclined plane is called the *wedge.* A wedge which is used for splitting wood has usually the form of a double inclined plane, as in Figure 46. The law of the wedge is the same as that of the inclined plane ; but since a wedge is usually driven by a blow instead of a force acting continuously, it is difficult to illustrate this law by experiments.

Fig. 46.

The wedge is especially useful when a large weight is to be raised through a very short distance. Thus, a tall chimney, the foundation of which has settled on one side, has been made upright again by driving wedges under that side. So, too, ships are often raised in docks by driving wedges under their keels. Cutting and piercing instruments, such as razors, knives, chisels, awls, pins, needles, and the like, are different forms of wedges.

82. *The Screw.* — In Figure 47 we have a machine called the *screw.* It is *a movable inclined plane, in which the inclined surface winds round a cylinder.* The cylinder is the *body* of the screw, and the inclined surface is its *thread.*

Fig. 47.

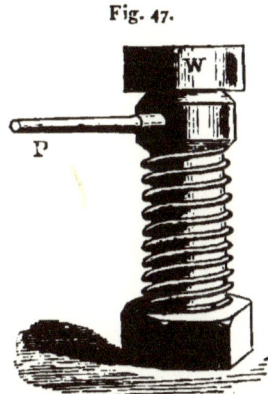

The screw usually turns in a block *N*, called the *nut.* Within the nut there are threads exactly corresponding to those on the screw. The threads of the screw move in the spaces between those of the nut.

The power is usually applied to the screw by means of a lever *P.* Sometimes the screw is fixed and the nut is movable, and sometimes the nut is fixed and the screw movable.

83. *The Endless Screw.* — In Figure 48 the thread of the screw works between the teeth of the wheel ; hence, if the screw is turned, the wheel must turn. Since as fast as the teeth at the left escape from the screw those on the right come up to it, the screw is acting upon the wheel continually ; hence this machine is called the *endless screw.*

Fig. 48.

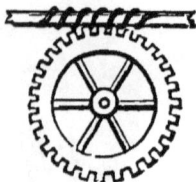

III.

PHYSICS.

I.

STATES OF MATTER.

A. THREE STATES OF MATTER.

84. *The Three States.* — Matter exists in three different *states*, known as the *solid*, the *liquid*, and the *gaseous*. Ice is a solid, water is a liquid, and steam and air are gases. While the substance of a body depends upon its atomic structure (2), the *state* of a body depends upon its *molecular* structure. Hence the state of matter is a *physical* condition, and changes of state are *physical* changes.

85. *Cohesion in the Different States of Matter.* — The different states of matter depend upon *the strength of the attraction of cohesion among the molecules.* This is comparatively strong in solids, very weak in liquids, and entirely wanting in gases.

The molecules of some solids are bound together much more firmly than those of others by cohesion ; but even when this bond is weakest, the molecules manifest a disposition to maintain their relative positions in the body, and the body to preserve its form. In liquids the bond of cohesion is so slight that the molecules manifest no disposition to maintain their relative positions in the body, nor does the body tend to preserve its form. Gases are not held together at all by cohesion, but only by gravity.

86. *Molecular Motion in the Different States of Matter.* —

The molecules are, undoubtedly, in incessant motion in every state of matter, but their freedom of motion is very different in the different states. In *solids*, the molecules, when left to themselves, have fixed positions within which they can move to a limited extent, but from which they can never escape. When left to themselves, the molecules of a solid never move around among themselves so as to change their relative positions. A molecule in the interior can never work its way to the surface, nor can one at the surface work its way into the interior.

In *liquids*, the molecules are all the time moving about among themselves in the interior of the mass with the utmost freedom. No molecule is confined within particular limits within the mass, but every molecule is continually moving to and fro in every direction throughout the entire mass. They, however, never escape from the influence of cohesion. So long as they are in the interior of the mass, the cohesion of the molecules on one side of them is exactly balanced by that of the molecules on the other side ; hence it does not interfere with the freedom of their motion. But as the molecules come to the surface, they experience only the pull of the molecules behind them, and this is usually sufficient to stop their outward motion and to cause them to return into the interior of the mass.

In *gases*, the molecules are moving without the slightest restraint from cohesion ; hence they move in straight lines. They are continually striking together and rebounding again, but after each rebound they move in straight lines till they encounter other molecules. There is no force acting within the mass of a gas which tends to check the motion of the molecules at any point ; hence gases do not, like liquids, tend to assume a definite surface.

87. *The Distances between the Molecules in the Different States of Matter.* — As a rule, the molecules are nearer together in solids than in liquids, and in liquids than in gases. The molecules of steam are about seventeen hundred times as far apart as those of water.

88. *Behavior of the Different States of Matter when Small Portions of each are placed in Empty Vessels.* — If a small portion of a *solid* is placed in an empty vessel, it will either not conform to the shape of the vessel at all, or, in the case of a soft

solid, only slowly and imperfectly. This is owing to the ten-dency of a solid to maintain its shape. If a small amount of a *liquid* is put into an empty vessel, it will conform at once and perfectly to the shape of the vessel, but it will not completely fill it. The liquid will sink to the lowest part of the vessel, and will be separated by a definite surface from the space in the upper part of the vessel. This is because the cohesion of the liquid checks the outward motion of the molecules, and so keeps them from moving away from the mass. If any portion of a *gas*, however small, is placed in an empty vessel, however large, the gas will completely fill the vessel. This is because there is nothing to check the outward motion of the molecules of the gas, save the walls of the vessel in which it is enclosed.

B. Fluids.

89. *Fluids.* — Owing to their freedom of molecular mo-tion, *liquids and gases have several characteristics in common.* They are, accordingly, often classed together as *fluids.* This appellation is derived from the readiness with which portions of each of these states of matter *flow* over or among each other.

90. *Pascal's Law.* — One of the most remarkable char-acteristics of a fluid is the way in which it transmits any pressure that is brought to bear on it. *If any pressure is brought to bear on any portion of the surface of a fluid which fills a closed vessel, a pressure just equal to it will be trans-mitted through the fluid to every equal portion of surface.* This law was enunciated by Pascal, and is known as *Pascal's law.*

The following experiment shows that pressure is trans-mitted in all directions by a fluid. A tube (Figure 49) is provided with a piston and fit-ted with a hollow globe, which is pierced with a number of orifices, arranged in a circle around it. Fill the globe and

Fig. 49.

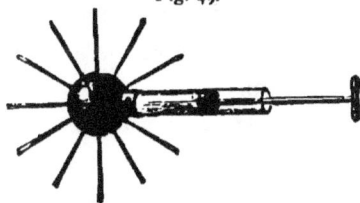

tube with water. If the piston is pushed in, the water spouts out of all the orifices, and not merely those opposite the piston.

Conceive a vessel of any form, in the sides of which are a number of cylindrical apertures, all of the same size, and closed with movable pistons, as shown at $A, B, C,$ D, and E (Figure 50). Suppose a pound of pressure brought to bear upon A. A pound of pressure will be transmitted to each of the other pistons in the direction of the arrows. If the piston B has only half the surface of A, it will receive only $\frac{1}{2}$ a pound of pressure ; if it has twice the surface of A, it will receive 2 pounds of pressure ; if it has three times the surface of A, it will receive 3 pounds of pressure ; etc. Hence, by means of a liquid, *a small pressure upon a small surface may be made to exert a great pressure upon a large surface.*

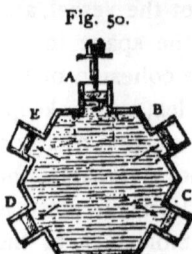

Fig. 50.

91. *The Hydraulic Press.* — In Figure 51 we have two cylinders, with a piston in each. Suppose that the surface of the larger piston is fifty times that of the smaller ; if the latter is pressed downward by a weight of one pound, an upward pressure of one pound will be brought to bear upon each portion of the surface of the large piston equal to that of the small piston. The whole upward pressure on the large piston will then be fifty times the downward pressure on the small one. If the surface of the larger piston had been one hundred times that of the smaller, one pound on the latter would have balanced one hundred on the former ; and so on.

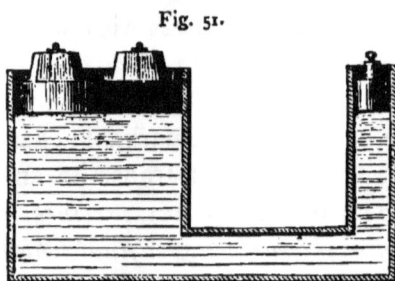

Fig. 51.

The *hydraulic press* is constructed on the principle just illustrated. One form of this press is shown in Figures 52 and 53. The two cylinders A and B are connected by the pipe *d*. The piston *a*, in the cylinder *A*, is worked by the handle *O*, and forces water into the large cylinder *B*, where it presses up the piston *C*. If the end of the piston *C* is 1000 times as large as that of the piston *a*, a pressure of 2 pounds on *a* would exert a pressure of 2000 pounds, or one ton, upon *C*. If a man, in working the

Fig. 52.

handle *O*, forces down the piston *a* with a pressure of 50 pounds, he would bring to bear upon *C* a pressure of 25 tons.

This press is used for pressing cotton, hay, cloth, etc., into bales; for extracting oil from seeds; for testing cannon, boilers, etc.; and for raising ships out of the water.

The *hydraulic jack* is a form of the hydraulic press, adapted to raising heavy weights.

92. *The Principle of Archimedes.* — *A body in a fluid is buoyed up by a force equal to the weight of the fluid it displaces.* This fact was discovered by Archimedes, and is therefore designated by his name.

This principle may be verified by the following experiment. A brass cylinder is constructed so as just to fill a cup. The cup and cylinder are hung from one pan of a balance (Figure 54) and counterpoised in the air by weights in the other pan.

Fig. 53.

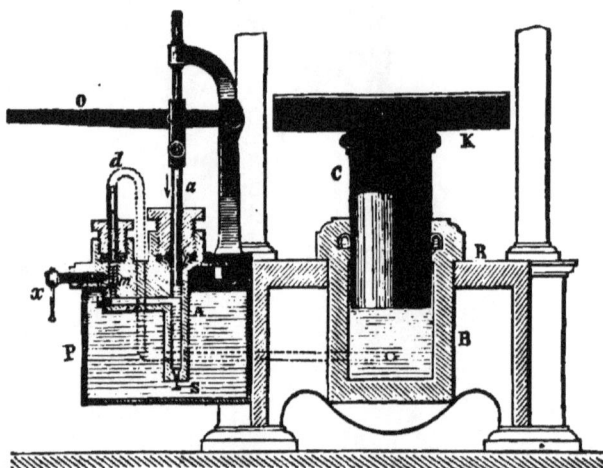

The cylinder is then allowed to hang in a vessel of water. The weights overbalance the cup and cylinder, showing that the water lifts the cylinder up. Equilibrium is restored by filling the cup with water. When the cup is full, the beam of the balance will be horizontal, and the cylinder will be completely in the water, showing that the cylinder is buoyed up by the water with a force equal to the weight of a cupful of water, or to *the weight of the water displaced* by the cylinder.

93. *Forces acting upon a Body immersed in a Fluid.* — Every body immersed in a fluid is subjected to two forces : one equal to *its own weight*, which tends to make the body

sink; the other equal to *the weight of the liquid displaced,* which tends to make the body *rise.*

When a body displaces *more than its own weight* of a

Fig. 54.

fluid, it will *rise* in that fluid ; when it displaces *less than its own weight,* it will *sink ;* and when it displaces *just its own weight,* it will *remain suspended* wherever it happens to be.

Fig. 55.

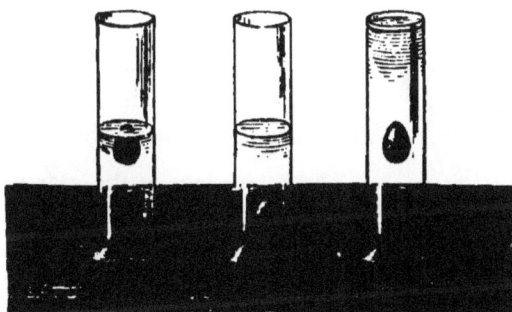

These three cases may be illustrated by putting an egg into salt and fresh water (Figure 55). When the egg is placed in salt water, it rises to the surface because it displaces more than its own weight of the brine. When it is put into the fresh water, it sinks to the bottom because it displaces less than its own weight of the water. When it is put into a proper mixture of fresh water and brine, it will remain suspended in the fluid, because it displaces just its own weight of the mixture.

Fig. 56.

94. *Floating Bodies.* — Every body floating in a fluid *displaces just its own weight* of the fluid. This is equally true of a ship floating in water, or a balloon floating in the air (Figure 56). The more heavily the ship is loaded, the deeper she sinks into the water. By throwing out the sand which is used as ballast, the balloon is made lighter, so as to displace more than its own weight of air. It then rises till it comes into more highly rarefied air, where it displaces just its own weight, when it again floats along at the same level. If some of the gas is allowed to escape, the balloon becomes less in bulk, and so displaces less than its own weight of air. It then sinks until it again displaces its own weight.

Fig. 57.

The appendage at the side of the balloon (Figure 56) is called a *parachute*, and can be used in descending from the balloon. It consists of a large circular piece of cloth (Figure 57) about 16 feet in diameter, which, by the resistance of the air, spreads out like a gigantic umbrella. In the centre there is an aperture, through which the air, compressed by the rapidity of the descent, makes its escape;

for otherwise oscillations might be produced, which would be dangerous to the aeronaut.

In Figure 56 the parachute is attached to the network of the balloon by means of a cord, which passes round a pulley, and is fixed at the other end to the boat. When the cord is cut the parachute sinks, at first very rapidly, but more slowly as it becomes distended, as represented in the figure.

95. *Method of finding the Specific Gravity of Solids and Liquids.* — To find the specific gravity (53) of a solid or liquid, it is necessary to *find the weight of a volume of water*

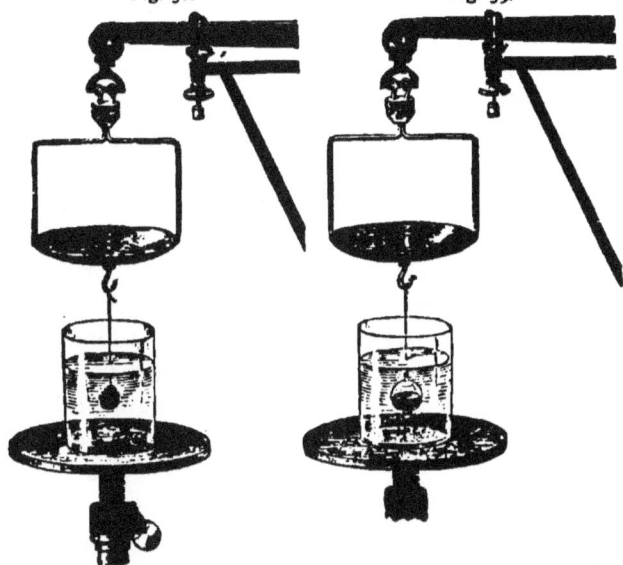

Fig. 58.   Fig. 59.

equal to that of a portion of the solid or liquid whose specific gravity is to be found. By means of the principle of Archimedes, the weight of this volume of water is easily found.

Suppose we wish to find the specific gravity of copper. Fasten the piece of copper to one pan of the balance by a fine thread (Figure 58), and counterpoise it in the air with weights in the other pan. Suppose it to weigh 125.35 grains. Then suspend it in a vessel of water and restore the equilibrium by placing weights in the pan supporting the copper. Suppose it to require 14.24 grains. This, according to the principle of

Archimedes, is the weight of the water displaced by the copper, or of a volume of water equal to that of the copper. The specific gravity of the copper is then $\frac{125 \cdot 3.5}{14 \cdot 21} = 8.8$.

When the body whose specific gravity we wish to find is *lighter than water*, we must fasten it to a heavy body to sink it. We then find, by the above method, the weight of the water displaced by the sinker alone, and by the sinker and light body together. The difference between the two will be the weight of the water displaced by the lighter body.

The specific gravity of a *liquid* may be found by the following method. A glass ball, weighted with mercury inside, is first accurately weighed in air. It is then immersed in a vessel of alcohol or other liquid under examination (Figure 59), and equilibrium is restored by adding weights to the pan from which the ball is suspended. Suppose 35.43 grains are required. This will be the weight of the ball's volume of alcohol. Next immerse the ball in water, and restore the equilibrium as before. Suppose it requires 44.28 grains this time. This will be the weight of the ball's volume of water. The specific gravity of alcohol will be $\frac{35.43}{44.28} = .8$.

96. *The Hydrometer.* — A *hydrometer* is an instrument *for finding the specific gravity of liquids.* Common forms of it are

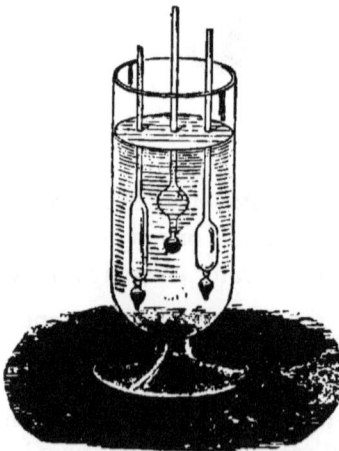

Fig. 60.

shown in Figure 60. They are weighted at the lower end with mercury to keep them in an upright position. The bulb above the mercury causes them to displace enough of a liquid to float in it. When put in a liquid they sink in it till they displace their own weight. The deeper they sink in a liquid, the less its specific gravity. Their stems are graduated in such a way that the number on the stem at the surface of the liquid indicates the specific gravity of the liquid. This is a convenient, but not very accurate method of ascertaining the specific gravity of a liquid.

C. GASES.

97. *Expansibility of Gases.* — One of the most marked characteristics of a gas is its *capacity for indefinite expansion.* The tendency of a gas to expand may be illustrated by means of an india-rubber bag partially filled with air, closed air-tight, and placed under the receiver of an air-pump. When the air is exhausted from the receiver, the bag fills out, as shown in Figure 61.

The tendency of a gas to expand is due to two facts, namely, that the molecules of a gas are *not held together by cohesion* (85), and that they are *moving rapidly in straight lines* (86). The condition of a gas in a closed vessel has been likened to that of a swarm of bees in a closed room, when all the bees are flying at random in straight lines. They would be constantly flying against one another and against the walls of the room. It has been calculated that the molecules of air are moving at the average rate of about 1600

Fig. 61.

feet a second. This velocity would be sufficient to carry a body in a vacuum some 40,000 feet, or about 7 miles high. Now the molecules of air in the rubber bag are all the time flying against one another and against the bag with this enormous velocity. They therefore tend to expand the bag. So long as there was air in the receiver outside the bag, the blows against the bag from within were met and balanced by an equal number of blows from without; but as the air was exhausted from the receiver, there were fewer and fewer blows upon the bag delivered by the molecules on the outside, and hence the bag began to yield to the more numerous blows from within.

98. *The Diffusion of Gases.* — When any two gases are brought into contact, they rapidly mix with each other.

This *mixture of gases when brought into contact* is called *diffusion.* It is due to the fact that the molecules are far apart and in constant motion. The molecules of the one gas quickly move into the spaces among the molecules of the other gas.

99. *The Expansive Power of a Gas increased by Heat.*— A bulb with a tube projecting from it is placed in a vessel of water so that the open end of the tube is under water, as shown in Figure 62. If the bulb is heated, the air in it will expand so as to drive out a portion of it through the water. *Heat always increases the expansive power of a gas.* This is

Fig. 62.

because heat causes the molecules to fly about with greater velocity, and therefore with greater energy.

100. *The Expansive Power of a Gas increased by an Increase of Pressure.* — An increase of pressure in a gas increases its expansive power. This is because the increased pressure *crowds the molecules nearer together,* so that there are more molecules in the same space to beat against the enclosure. In the cylinder of the steam-engine the steam is kept at a high temperature and under great pressure.

101. *The Three Laws of Gases.* — *Equal volumes of all gases, at the same temperature and under the same pressure, contain the same number of molecules.* This is *Avogadro's law.*

The volume of a confined mass of gas varies inversely as the pressure to which it is exposed. The less the pressure the greater the volume, and the greater the pressure the less the volume. This is *Mariotte's law.* This law might be stated thus : the number of molecules of a gas in a given space, and the expansive power of the gas, vary directly as the pressure to which the gas is exposed.

*The volume of a gas under constant pressure varies directly
as the absolute temperature of the gas.* This is *Charles's
law.*

By *absolute temperature* is meant temperature *measured from
a point 459° below the ordinary zero.* The temperature indicated by an ordinary thermometer may be converted into absolute temperature by adding 459° to it. Thus, a temperature of
70° on our scale would be a temperature of $70° + 459° = 529°$ on
the absolute scale. A temperature of $-15°$ on our scale would
be a temperature of $459° + (-15°) = 444°$ on the absolute scale.

Fig. 63.

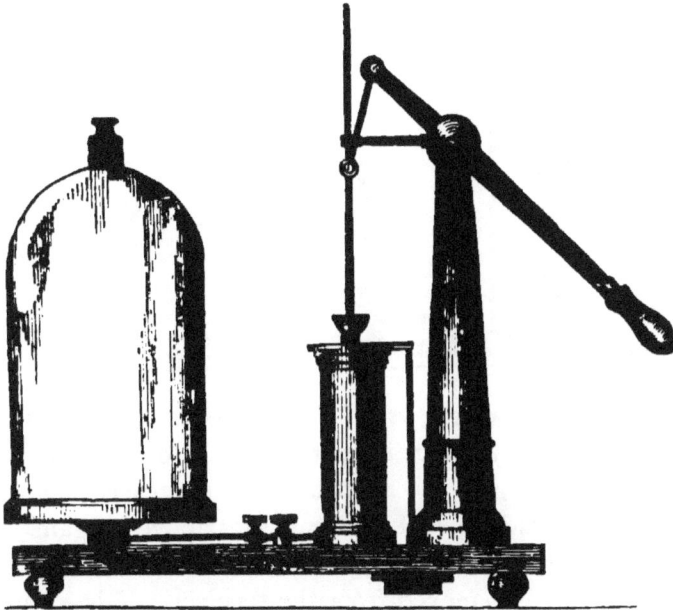

102. *The Air-Pump.* — The essential parts of an *air-
pump* are shown in Figures 63 and 64. There is a flat
plate for holding the *receiver E*, called the *pump-plate.* It
is ground perfectly flat, so that an *air-tight* joint is formed
between it and the receiver when the latter is placed upon
it. A tube connects the pump-plate with the *cylinder*, in
which a piston is moved up and down by means of the

5

handle. There is a little valve S in the piston, pressed down by a spiral spring above it. There is also a valve S' at the bottom of the barrel, fastened to a rod which passes through the piston in such a way that the valve is opened when the piston rises, and closed when the piston is pushed down, by the friction of the rod against the piston. When the piston is drawn up the valve in the piston is closed, and no air can pass from above the piston into the space below it. At the same time S' at the bottom of the barrel is opened, and the expansive force of the air in the receiver E causes some of the air to pass out through the tube into the barrel below the piston. When the pis-

Fig. 64.

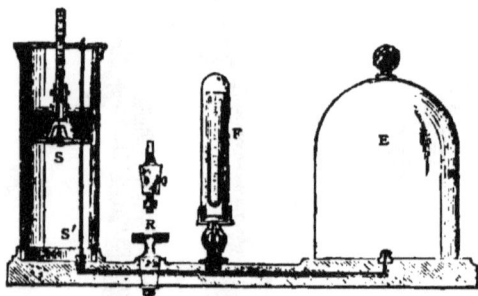

ton is pushed down the valve S' is closed by the friction of the rod, and the valve S is opened by the expansive force of the air below it as the air becomes compressed, and the air in the barrel below the piston passes above it again. In this way, every time the piston is moved up and down, a part of the air is removed from the receiver. F is a gauge for showing the extent of the exhaustion ; R is a cock, by means of which the receiver and the barrel may be put into communication with each other, or either may be shut off from the other, and be put into communication with the external air.

There are many different forms of air-pumps ; but with none of the ordinary pumps is it possible to obtain perfect exhaustion.

The air becomes finally so attenuated as not to have sufficient expansive force to open the valve.

103. *Pressure of the Air.* — The pressure of the air may be illustrated by the following experiments. Place a small bell-jar, open at both ends, on the plate of the air-pump, and cover the top of the jar with the palm of the hand. When the air is exhausted from the jar, the hand is pressed firmly down upon the mouth of the jar. This is an illustration of the *downward pressure* of the air. It was not perceived at first, because the downward pressure of the air upon the hand was balanced by the upward pressure of the air within the jar.

The *weight-lifter* (Figure 65) serves to illustrate the *upward pressure* of the air. It consists of a cylinder of glass or metal, *A B*, with a piston moving up and down in it, air-tight. The cylinder is closed at the top by a plate *C*, to which may be screwed a tube to connect the cylinder with the air-pump. The cylinder is open at the bottom, and a heavy weight is fastened with a strap to the piston. If the air is exhausted from the cylinder above the piston, the piston and weight are raised by the upward pressure of the air acting upon the bottom of the piston.

Fig. 65.

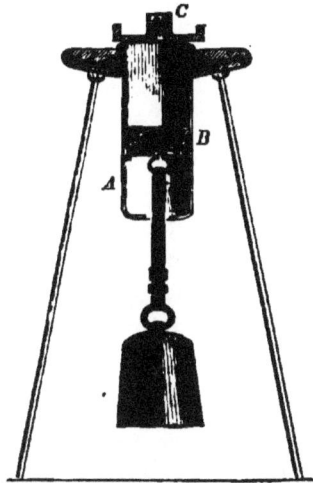

Figures 66 and 67 represent two brass hemispheres, some four inches in diameter, the edges of which are made to fit tightly together. The whole can be screwed to the air-pump by means of the stop-cock at the bottom. While the hemispheres contain air they can be separated with ease, since the outward pressure is just balanced by the inward pressure; but when the air within is pumped out, it is very hard to pull them apart. Since it is equally difficult to do this in whatever position the hemispheres are held, the experiment shows that the air *presses in all directions.*

This piece of apparatus is called the *Magdeburg hemispheres*, from Otto von Guericke, of Magdeburg, by whom it was invented.

The pressure of the air at the level of the sea is about
15 *pounds to a square inch,* or a ton to the square foot.

Fig. 66.

The surface of the body
of a man of middle size is
about 16 square feet ; the
pressure, therefore, which a
man supports on the surface
of his body is 35,560 pounds,
or nearly 16 tons. Such
enormous pressure might
seem impossible to be borne;
but it must be remembered
that, in all directions, there
are equal and contrary pressures which counterbalance
one another. It might also
be supposed that the effect
of this force, acting in all
directions, would be to press the body together
and crush it. But the solid parts of the skeleton
could resist a far greater pressure ; and the cavities of the body
are filled with air or liquids which exert a pressure outward
equal to that of the external air. When the external pressure
is removed from any part of the body, either by means of a
cupping vessel or by the air-pump, the pressure from within
is seen by the distension of the surface.

Fig. 67.

104. *The Pressure of the Air decreases as we ascend above
the Level of the Sea.* — The pressure of the air at the level
of the sea is due to the *downward pressure of all the layers
of air above,* transmitted throughout the mass below according to Pascal's law (90).

Each layer of molecules of air is pulled downward by gravity,
and transmits this pressure to all the layers below. Hence the
pressure of a gas increases with the depth. It, however, *increases more rapidly than the depth.* For, gases being compressible, as we descend in a gas the molecules are crowded more
closely together, so that there are more molecules exerting pres-

sure in each layer, and there are more layers in any given difference of depth.

D. LIQUIDS.

105. *Compressibility of Liquids.* — For a long time it was thought that liquids were entirely incompressible. In the year 1661 some academicians of Florence, wishing to find whether water was compressible, filled a thin globe of gold with that liquid, and, after closing the orifice perfectly tight, subjected the globe to great pressure, with a view of altering its form, knowing that any alteration of form would occasion a diminution of capacity. They failed to compress the water, but discovered the porosity of gold, for the water forced its way through the pores of the globe, and stood on the outside like dew.

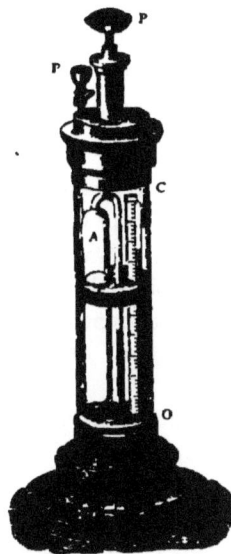

Fig. 68.

In more recent times it has been shown that liquids are *slightly compressible.*

The apparatus for measuring the compressibility of a liquid is shown in Figure 68. It consists of a strong glass cylinder enclosing a long glass bulb *A*, from which proceeds a fine bent tube, with its end dipping under the mercury in the bottom of the cylinder at *O*. The liquid to be tested is introduced into the bulb *A* so as to fill both it and the tube. The cylinder is then filled with water through the funnel *R*, and pressure applied by means of the thumb-screw *P*, which forces a piston down upon the water. The rise of the mercury in the fine tube shows the amount of the compression of the liquid in the bulb. For a pressure of one atmosphere, or 15 pounds to the square inch, the volume of water is diminished about 5 parts in 100,000. At the depth of a mile, the volume of sea-water is diminished 1 part in 130.

In liquids, as in gases, *elasticity is developed only by compression*, but their elasticity is *perfect*. No matter to what pressure a liquid has been subjected, it will return to exactly its original volume as soon as the pressure is removed.

106. *The Tendency of Liquids to assume a Globular Form.* — When left to itself, a liquid always *assumes a globular form*. This is because all the molecules, as they work their way through the mass, are stopped by the force of gravity and cohesion at the same distance from the centre of the mass.

The tendency of the molecules of liquids to collect into spheres may be shown by the following experiment. Prepare a mixture of water and alcohol which shall be just as heavy as sweet oil, bulk for bulk, and introduce some of the oil carefully into the centre of this mixture by means of a dropping-tube; the oil will neither rise nor sink, but gather into a beautiful sphere.

Rain-drops, dew-drops, and the manufacture of shot illustrate this tendency of the molecules of liquids. In the manufacture of shot, melted lead is poured through a sieve at the top of a very high tower, and the drops in falling take the form of spheres, which become solid before they reach the bottom.

107. *The Free Surface of a Liquid at Rest is a Level Surface.* — A *level* surface is one *along which gravity does not tend to produce any motion*. Gravity always acts perpendicularly to such a surface, and hence there can be no component of gravity which would tend to produce motion along that surface.

The surface of a liquid at rest must be a level surface, else gravity would tend to move the liquid along the surface, and the liquid could not remain at rest.

108. *The Downward Pressure of a Liquid due to Gravity is proportioned to the Depth.* — Since the downward pressure of a liquid due to gravity at any point is the pressure that has been transmitted to that point by the layers

of molecules above, the pressure at that point will be *proportional to the number of layers of molecules above the point;* and since liquids are practically incompressible, the number of layers of molecules will be *proportional to the depth.*

The amount of pressure transmitted to the layers below by any layer of molecules is entirely independent of the extent of the layer. For if the upper layer consisted of a single molecule, it would exert the pressure of a molecule upon the surface of a molecule, and that pressure would be transmitted to every equal surface below. If the upper layer consisted of 5 molecules, they would exert a pressure of 5 molecules upon a surface of 5 mole-

Fig. 69.

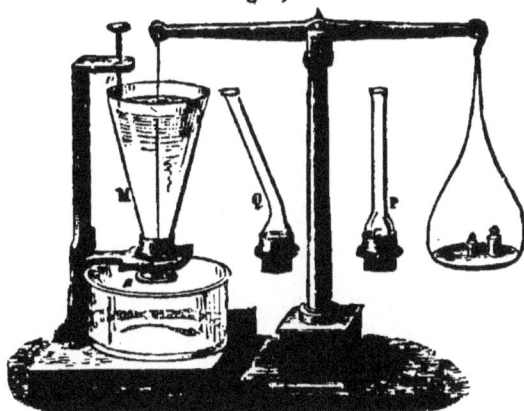

cules, which would be the pressure of one molecule to the surface of one molecule as before. Hence the pressure at any point in a vessel containing a liquid does not depend at all upon the size and shape of the vessel, but simply upon the depth of the point below the surface.

109. *Pascal's Vessels.* — The fact that the pressure of a liquid upon a given surface *depends upon the depth of the liquid only,* and not upon the size or shape of the vessel which contains the liquid, may be illustrated by means of *Pascal's vessels* (Figure 69). The vessels *M, P,* and *Q* may in turn be screwed into the plate *c.* A disc *a* suspended from one end of the beam of a balance with a thread, and held up by weights at the other

end of the beam, serves as the bottom of the vessel, which it closes water-tight. Water is poured carefully into the vessel *M* till its depth is just sufficient to displace the plate *a*, and the height of the water is marked by the point *o*. *M* is then removed, and *P* and *Q* are in turn put into its place. It will be found that each will have to be filled to exactly the same height to displace the plate *a*.

It follows from the above that *a very small quantity of water can produce very great pressure*. Let us imagine a cask, for example, filled with water, and having a long narrow tube tightly fitted into its top. If water is poured into the tube, there will be a pressure on the bottom of the cask equal to the weight of a column of water whose base is the bottom itself, and whose height is equal to that of the water in the tube. The pressure may be made as great as we please; by means of a mere thread of water forty feet high, Pascal succeeded in bursting a very solidly constructed cask.

110. *The Upward Pressure of a Liquid.* — The downward pressure of a liquid at any point *must be balanced by an equal upward pressure*, according to the law that action and reaction are always equal and opposite (30).

The following experiment (Figure 70) serves to show the upward pressure of liquids. A large open glass tube *A*, one end of which is ground, is fitted with a ground-glass disc *O*, or still better with a thin card or piece of mica, the weight of which may be neglected. To this is attached a string *C*, by which it can be held against the bottom of the tube. If the whole is then immersed in water, the disc does not fall, although no longer held by the string; it is consequently kept in its position by the upward pressure of the water. If water is now slowly poured into the tube, the disc will sink only when the height of the water inside the tube is equal to the height outside.

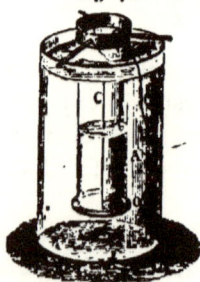
Fig. 70.

111. *The Pressures of different Liquids at the same Depth*

are proportional to their Densities. — The pressure at the same depth would be about 12½ times as great in mercury as in water, and about .8 as great in alcohol as in water. This is owing to the fact that, mercury being about 12½ times as dense as water, each layer of mercury would transmit downward 12½ times as much pressure as a layer of the same thickness of water; and a layer of alcohol .8 times as much.

112. *The Pressure is the same at every Point in a Horizontal Layer of a Liquid at Rest.* — Owing to the extreme mobility of liquids, it would be impossible for a liquid to remain at rest if at any point in it the pressures acting upon that point from all directions were not equal or balanced.

If the upward or downward pressure at any point were not balanced, a particle at that point would tend to move up or down as the case might be. If the pressure were not the same throughout a horizontal layer, there would be some point in the horizontal layer where the horizontal pressures to the right and left would not be balanced, and a particle at that point would move in the direction in which it was urged by the greater pressure ; that is, the liquid would not be at rest. This is true of all fluids, both liquids and gases.

Any disturbance of the equilibrium of pressure in horizontal layers gives rise to currents which will flow towards the region of low pressure till the equilibrium is restored.

113. *Rise of Liquids in Communicating Vessels.* — When a liquid is contained in vessels which communicate with each other and is at rest, it will be found to stand *at the same height in all the vessels*, whatever may be their size or shape.

Thus, in Figure 71, the water stands at the same height in all the tubes as in the large vessel. If one of the tubes is cut off below the level of the water in the other vessels, and drawn out to a narrow mouth, the liquid will spout out of this tube nearly to the height of the liquid in the others. The rise of a liquid

to the same height in a series of communicating vessels is due
to the fact that when a liquid is at rest the pressure must be

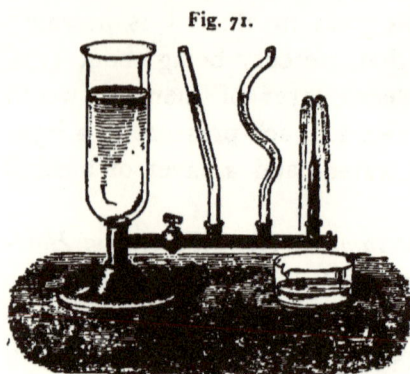

Fig. 71.

the same throughout each
horizontal layer. Each
horizontal layer of the
water taken through all
the vessels must be the
same distance below the
free surface of the liquid
in each vessel. Hence
these free surfaces must
be in the same horizontal
line, or at the same level.

The tendency of liquids
to find their own level is very important, and of continual application. When any system of pipes, however complicated, is
connected with a reservoir, the water will rise in every pipe to
the level of the water in the reservoir.

114. *Springs and Artesian Wells.* — All natural collections
of water illustrate the tendency of a liquid to find its level.
Thus, the Great Lakes of North America may be regarded as a
number of vessels connected together, and hence the waters
tend to maintain the same level in all. The same is true of the
source of a river and the sea, the bed of the river connecting
the two like a pipe.

Springs illustrate the same fact. The earth is composed of
layers, or *strata*, of two kinds : those through which water can
pass, as sand and gravel ; and those through which it cannot
pass, as clay. The rain which falls on high ground sinks
through the soil until it reaches a layer of this latter kind, and
along this it runs until it finds some opening through which it
flows as a spring.

It is the same with *Artesian wells.* These wells derive their
name from the province of Artois in France, the first part of
Europe where they became common. It would seem, however,
that wells of the same kind were made in China and Egypt,
many centuries earlier.

In Figure 72 suppose *A B* and *C D* to be two strata of clay,
and *K′ K′* to be a stratum of sand or gravel between them. The

rain falling on the hills on either side will filter down through this sand or gravel, and collect in the hollow between the two strata of clay, which prevent its escape. If now a hole is bored

Fig. 72.

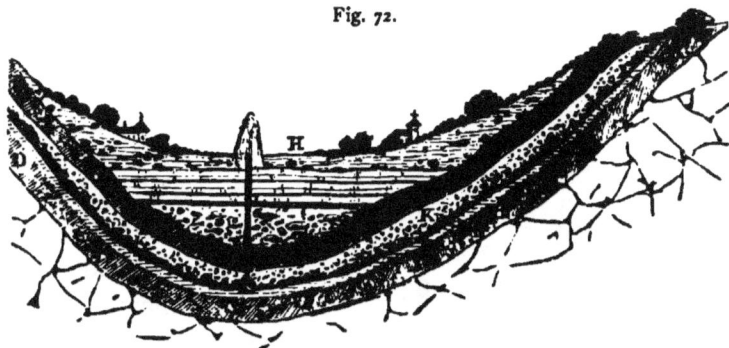

down to $K'K$, the water, striving to regain its level, will rise to the surface at H, or spout out to a considerable height above it.

Sometimes the water between two such impervious strata makes its way to the surface through some fissure in the upper stratum, constituting a deep-seated spring.

115. *The Spirit-Level.* — The *spirit-level* consists of a closed glass tube, AB (Figure 73), with a slight upward curvature. It is filled with spirit, except a bubble of air which tends to rise to the highest part of the tube. It

Fig. 73.

is set in a case CD, and when this is placed on a perfectly level surface the bubble is exactly in the middle of the tube, as in the figure.

116. *Rise of two Different Liquids in Communicating Vessels.* — If into one of two communicating tubes (Figure 74) we pour any liquid, as mercury, it will rise to the same height in both branches. If now we pour water into one of the tubes, the mercury will rise somewhat in the other, but not nearly so high as the water. The height of the two liquids above the surface of separation will be in the inverse ratio of the densities of the liquids. This is because the pressures of the two liquids at the surface of separation must be equal, so as to balance each other. Now the downward pressure of the water at the surface of the

mercury is due to the depth of the water above it, and the upward pressure of the mercury at the same point is due to

Fig. 74.

the depth of the mercury above the level of this surface in the other tube ; and to have these pressures equal, *the depths must be in the inverse ratio of the densities of the liquids.*

117. *Capillarity.*— The rise of liquids in communicating vessels is modified in a remarkable manner when any of the vessels are *of small diameter.* Such narrow vessels and fine tubes are called *capillary,* from the Latin *capillus*, a hair ; and their action upon the rise of liquids within them is known as *capillary action.*

This action is not, however, confined to the cases of fine tubes; but when the containing vessel is wide, the action extends only a short distance from the sides of the vessel. The free surface of a liquid in a wide vessel is not horizontal in the neighborhood of the sides of the vessel, but presents a decided curvature. When the liquid *wets* the vessel, as in the case of water in a glass vessel (Figure 75), the surface of the liquid near the sides is *concave.* When the liquid *does not wet* the vessel, as in the case of mercury in a glass vessel (Figure 76), the surface near the sides is *convex.*

When a narrow tube of glass is plunged into water or any other liquid that will *wet* it (Figure 77), the liquid rises *higher within the tube than on the outside,* and the column of liquid within the tube will be *concave* at the top. In this case there is a *capillary ascension* which varies in amount with the diameter of the tube and the nature of the liquid. The finer the tube, the higher the liquid will rise in it. If a glass tube is plunged in mercury, which does not wet it, the mercury will *fall within the tube below the level outside* (Figure 78), and the top of the

Fig. 75.   Fig. 76.   Fig. 77.   Fig. 78.

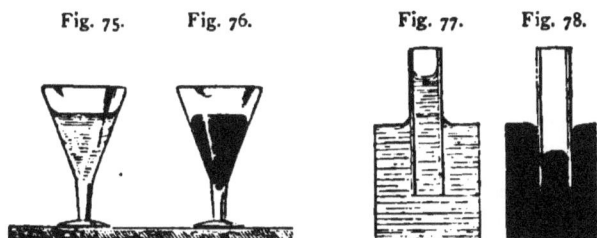

column of mercury within the tube will have a *convex* surface. In this case there is a *capillary depression.* The finer the tube, the greater the depression.

If we take two bent tubes, each having one branch of considerable diameter, and the other extremely narrow, and pour water into one of the tubes, and mercury into the other, the water will stand higher in the capillary than in the principal branch, and the mercury will stand lower in the capillary branch (Figure 79). The free surface will be concave in both branches in the case of water, and convex in the case of mercury. Capillary action is manifested *whenever the surface of a liquid comes in contact with a solid.* If a clean glass plate is dipped into water, the water will rise a little on each side of the plate. If the same plate is dipped into mercury, the mercury will be depressed a little on each side of the plate.

Fig. 79.

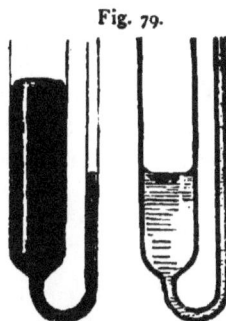

118. *Illustrations of Capillarity.* — A lamp-wick is full of tubes and pores, and capillary force draws the oil up through these to the top of the wick, where it is burned. When one end of a cloth is put into water, capillary force draws the water into the tubes and pores of the cloth, and the whole soon becomes wet. In the same way any other porous substance soon

becomes wet throughout, if a corner of it is put into water. Blotting-paper is full of pores into which the capillary force draws the ink. The use of a.towel for wiping anything which is wet depends on the same principle.

119. *Strength of the Capillary Force.* — It is well known that when a piece of cloth is wet, it is almost, if not quite, impossible to wring or squeeze it dry. This shows that the capillary force which holds the water in the pores of the cloth is very strong. Some solids, as wood, swell on becoming wet. If holes are drilled into a granite rock, and dry wooden plugs driven into them, and water is then poured over the ends of the plugs, the capillary force draws the water into the wood, which swells and splits the rock. This is a striking illustration of the strength of the capillary force.

120. *Capillary Force never causes a Liquid to flow through a Tube.* — If a glass tube is so fine that the capillary force will draw water into it to the height of two inches, and the tube is then lowered so that not more than half an inch shall be above the surface of the water, the water will not overflow the tube. If, however, the water is removed as soon as it comes to the top, more will rise in the tube to take its place.

When a lamp is burning, the oil is passing up continually through the wick, because it is burned as soon as it reaches the top ; but when the lamp is not burning, the oil does not overflow the wick.

Fig. 80.   Fig. 81.

121. *Heavy Bodies floating on Water by Capillary Action.* — According to the principle of Archimedes (92), a body cannot float on a liquid unless it is less dense than the liquid. This seems to be contradicted by certain well-known facts. Small steel needles will float on water when placed carefully on the surface (Figure 80). Some insects walk on water (Figure 81),

and many heavy bodies can, if sufficiently minute, float on the surface of water. In all these cases the bodies are not *wet* by the liquid, and consequently depressions are formed around them by capillary action, as shown in Figure 82. The liquid displaced by one of these bodies is really equal to that which would fill the whole depression, or the space below the dotted line *C D* (Figure 82), and this liquid would in every case be equal to the weight of the floating body.

Fig. 82.

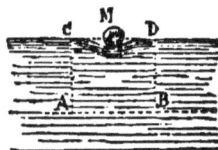

122. *Rise of Liquids in Exhausted Tubes.* — Since the atmosphere presses 15 pounds to the square inch upon the surface of a liquid, *if this pressure is removed or lessened* at any point on the surface, *the liquid will tend to rise* at that point. If a long glass tube, open at both ends, is connected at the top by means of a rubber tube with an air-pump, and is held upright with its lower end under the surface of mercury, when the pump is worked the mercury will begin to rise in the tube, and it will rise higher and higher as the exhaustion continues. If the air could be entirely exhausted, the mercury would rise about 30 inches in the tube. Under similar circumstances water would rise about 33 feet high. In each case the liquid would rise in the tube till the pressure within the tube at a level with the surface of the liquid outside was equal to the pressure of the air on the surface of the liquid, or about 15 pounds to the square inch. The *height* to which different liquids will rise in exhausted tubes will be *in the inverse ratio of the densities of the liquids.*

In drinking lemonade through a straw, the air is first drawn out of the straw by the mouth, and the liquid is forced up through the straw by the pressure of the air on the surface. When a jar is filled with a liquid and then inverted with its mouth under the same liquid in a vessel, the pressure of the air on the surface of the liquid in the vessel will keep the liquid up in the jar.

That it is the pressure of the atmosphere on the surface of the liquid in the vessel that keeps the liquid up in the jar may be shown by the following experiment. Fill a jar with mercury, invert it, and place its mouth under some mercury in a dish. Place the jar thus inverted in the dish of mercury under the receiver of an air-pump, and exhaust the air. As the exhaustion proceeds, and the pressure of the air upon the surface of the mercury becomes less and less, the mercury falls in the jar.

123. *The Fountain in Vacuo.* — This apparatus is an illustration of the tendency of liquids to rise in exhausted vessels

Fig. 83.

(Figure 83). It consists of a bell-jar, provided with a tube and stopcock at the bottom. The bell-jar is first exhausted by means of the air-pump. The stopcock is then closed, and the bell-jar is removed to a vessel of water. After the end of the tube has been placed under water the stopcock is again opened. The pressure of the air on the surface of the water in the vessel drives the water up in the bell-jar in a jet so as to form a beautiful fountain.

124. *Torricelli's Experiment.* — Torricelli took a glass tube somewhat more than 30 inches long and closed at

one end, and filled it with mercury. He then closed the tube with his thumb, and inverted it in a dish of mercury (Figure 84). On opening the tube under the mercury, he found that the mercury fell in the tube till the top of the column *A* stood about 30 inches above the surface of the mercury in the dish. Such a tube is called a *Torricellian tube*, and the space above the column of mercury in the tube is called a *Torricellian vacuum*.

Fig. 84.

125. *Pascal's Experiment.* — Pascal had a Torricellian tube taken from the bottom to the top of a mountain, and found that the column of mercury in the tube fell as the ascent progressed. He therefore concluded that the mercury was kept up in the tube by *the pressure of the atmosphere on the surface of the mercury in the vessel*, since the pressure would necessarily become less and less as we ascend from the level of the sea.

126. *The Barometer.* — The *barometer* is an instrument for *measuring the pressure of the atmosphere.* It is a Torricellian tube furnished with a convenient case (Figure 85). The vessel of mercury at the bottom must be constructed so as to prevent the spilling of the mercury in transportation, and so as to allow the atmosphere to act freely upon the mercury.

6

Fig. 85.

127. *Use of the Barometer in measuring the Height of Mountains.* — One of the chief uses of the barometer is to *measure the height of mountains.* It has already been stated that the atmospheric pressure is less as the height above the earth is greater. When we have found at what rate it diminishes, we can readily find the height of mountains by means of the barometer. We have to find the difference between the readings of the barometer at the level of the sea and at the top of the mountain. This shows how much the pressure has diminished, and from this we can find the height of the mountain.

If the pressure of the atmosphere decreased uniformly as we ascend, it would be very easy to find the elevation of a place by means of a barometer. But, owing to the variations in the density of the air as we ascend, the pressure changes according to a complicated law ; and this complicates the formula for finding the exact elevation of a place from the readings of the barometer. As a rough rule, it may be stated that the barometer falls one inch for every 900 feet of ascent.

128. *The Suction-Pump.* — The *suction-pump* consists of a cylinder, or barrel, at the top of a pipe *A* (Figure 86), communicating with the water in the well or cistern. A piston *P* is moved up and down in the barrel by means of the handle *B*. There is a valve *S* at the top of the pipe *A*, and another valve *O* in the piston. Both valves open upwards. The pump first exhausts the air from the pipe. As the air is exhausted, the water is driven up through the pipe and finally into the pump-barrel by the pressure of the air on the surface of the water in the cistern. Every time the piston is pushed down the valve *S* closes, and keeps the water in the barrel from passing back into the cistern; at the same time the valve in the piston opens, and allows the water below it to pass above it. When the piston is raised, the valve *O* closes, and keeps the water above it from passing below it; at the same time the valve *S* is forced open by the pressure from below, and the water rushes up through it to fill the barrel behind the piston. As the piston is raised, the water above the piston passes out by the discharge-

Fig. 86. Fig. 87.

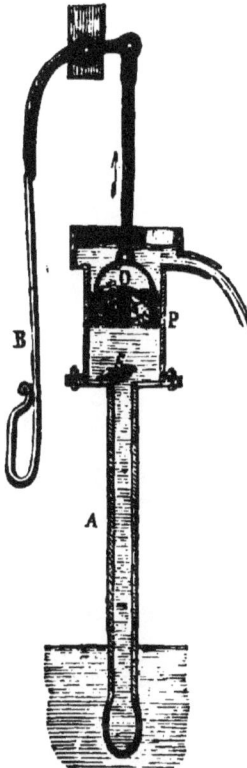

pipe at the top of the barrel. With this pump the water is raised into the barrel *by atmospheric pressure*, and is then lifted out of the barrel by the piston. Hence with the suction-pump *water can be raised only about 30 feet high*.

129. *The Force-Pump.* — The simple *force-pump* is shown

in Figure 87. The piston P is solid. The discharge-pipe D communicates with the bottom of the cylinder, and has a valve O in it opening upward. There is also a valve S in the bottom of the barrel, also opening upward. When the plunger is raised the valve O closes, and the water rushes into the cylinder through the valve S; when the plunger is pressed down, the valve S closes, and the water is forced out through the valve O into the discharge-pipe. The only limit to the height to which water may be raised by means of this pump is that of the power used and of the strength of the pump.

Fig. 88.	Fig. 89.	Fig. 90.

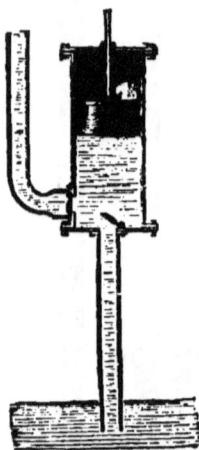  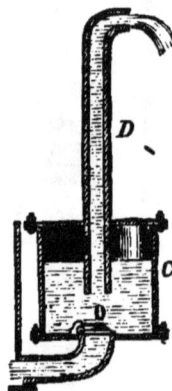

The *force-pump* and the *suction-pump* may be *combined*, as shown in Figures 88 and 89; that is to say, the cylinder of the force-pump may be at the top of a pipe about 30 feet above the surface of the water to be raised.

130. *The Air-Chamber.* — The *air-chamber* is a device by which the water from a force-pump may be made to escape in *a continuous and forcible stream*. It consists of an air-tight box C above the valve O in the discharge-pipe (Figures 87 and 90). The pipe D passes nearly to the bottom of the chamber. When the pump is working,

the water is forced into the air-chamber through the valve
O. As soon as the end of the pipe *D* is covered, the air
in the upper part of the chamber begins to be compressed.
The compression *increases the elastic force of the air*, and
causes it to press steadily and powerfully on the surface
of the water, forcing the liquid out through the pipe *D* in a
steady stream. If *D* ends in a narrow nozzle, the water
will be obliged to pass through it very rapidly to escape
from the chamber as rapidly as it is pumped into it. In
this way a stream may be obtained of sufficient force to
be thrown a great distance, as in the *fire-engine*.

131. *The Siphon.* — The *siphon* is used for transferring
liquids from one vessel to another. It consists of *a bent
tube with arms of unequal length* (Figure 91). The
air must be removed from the tube in the first place,
either by applying the mouth to the end *B*, after the other
arm of the siphon has been introduced into the vessel of
water, or by filling the siphon with water before it is
placed in the vessel. The water will flow through the
siphon from *C* to *B* until
the vessel is emptied, or
until the level of the water
falls below the mouth of the
arm in the vessel.

Fig. 91.

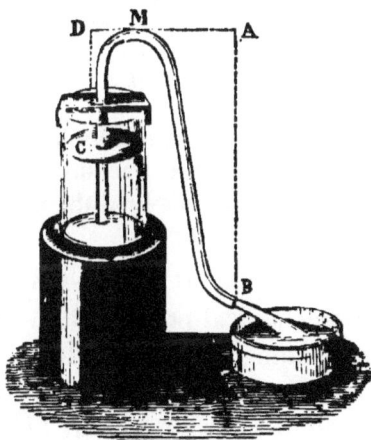

The flow of the liquid
through the siphon seems opposed to the well-known fact
that water will not run up hill.
But notwithstanding this seem-
ing inconsistency, it will be
seen that the water is flowing
from a higher level *C* to a
lower level *B*. If we consider
a layer of water in the siphon
at *M*, we see that the force which acts upon it from left to right
is equal to the pressure of the atmosphere minus the pressure of

the water in the tube from M to C, whose depth is $D\,C;$ and the pressure which acts upon it from right to left is equal to the pressure of the atmosphere minus the pressure of the water in the tube from M to B, whose depth is $A\,B$. Since $A\,B$ is greater than $D\,C$, the pressure at M towards the right will be greater than that towards the left. Consequently the water at M moves on towards B, and as it moves away more water is driven up into the arm $C\,M$ to take its place by the pressure of the atmosphere on the surface of the water in the vessel. No liquid will flow through a siphon unless the atmospheric pressure is sufficient to raise it to the bend of the tube.

132. *Tantalus's Cup.* — This is *a glass cup, with a siphon tube passing through the bottom*, as shown in Figure 92. If water is poured into the cup, it will rise both inside and outside the siphon until it has reached the top of the tube, when it will begin to flow out. If the water runs into the cup less rapidly

Fig. 92.

than the siphon carries it out, it will sink in the cup until the shorter arm no longer dips into the liquid, and the flow from the siphon ceases. The cup will then fill, as before ; and so on.

In many places there are *springs* which flow at intervals, like the siphon in this experiment, and whose action may be explained in the same way. A cavity under ground (Figure 93) may be gradually filled with water by springs, and then emptied through an opening which forms *a natural siphon*. In some cases of this kind the flow stops and begins again several times in an hour.

133. *Water-Wheels.* — One of the most important sources of mechanical power is that of *falling water*. The falling or running water is made to turn a wheel called a *water-wheel;* and this wheel, by means of bands or gearing, is made to work almost any kind of machinery.

Water-wheels are of various forms. Some turn on an upright axis, and others on a horizontal axis. The latter are called *vertical water-wheels*, and the former *horizontal water-wheels*.

Fig. 93.

One of the most common of vertical water-wheels is represented in Figure 94. It consists of a series of boxes, or *buckets*, arranged on the outside of a wheel or cylinder. Water is allowed to flow into these buckets on one side of the wheel, and by its weight causes the wheel to turn. The buckets are so constructed that they hold water as long as possible while they are going down, but allow it all to run out before they begin to rise on the other side.

Fig. 94.

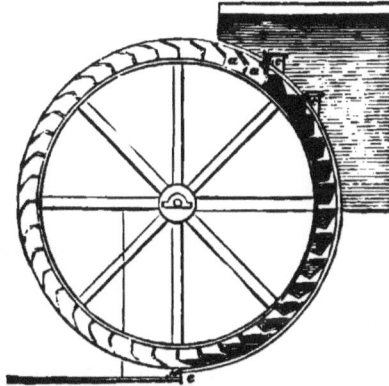

A wheel like this is called a *breast-wheel*.

The *overshot* wheel is similar to the breast-wheel in all respects, except that the water is led over the top of the wheel, and poured into the buckets on the other side.

The *undershot* wheel has boards projecting from its circumference, like the paddle-wheel of a steamboat. The water runs under the wheel, and turns it by the force of the current pressing against the boards.

134. *The Hydraulic Tourniquet.* — If a vessel (Figure 95), having a spout and faucet on one side, is filled with water

Fig. 95.

and floated in a dish on water so as to move easily, and the faucet is then opened so as to allow the water to escape, the vessel will begin to move backward. This is due to *the reaction of the water* against the back of the vessel. While the faucet was closed, the pressure of the water against the front of the vessel at the orifice balanced the pressure of the water against the back of the vessel at the same point. But when the faucet is open, there is no pressure against the front of the vessel to balance the reaction of the water against the back; hence the backward motion of the vessel while the water is escaping.

The *hydraulic tourniquet* (Figure 96) consists of a vessel

Fig. 96.

capable of turning on a vertical axis. Two tubes project from the bottom of the vessel in opposite directions. The ends of these tubes are open, and are bent round in opposite directions. As the water escapes from these tubes, its reaction against the parts of the tubes opposite the openings causes the apparatus to rotate rapidly.

135. *Turbine Wheel.* — One form of the *turbine wheel* is shown in Figure 97. This wheel turns in a horizontal plane. The buckets are placed in the outer part of the wheel, which is free to turn on a vertical axis. The curved partitions, or *guides*, within the wheel are stationary. These partitions are placed at the bottom of a long cylinder, into which the water is admitted by the pipe. The partitions are curved, so as to direct the water against the buckets at the most advantageous angle. The water is discharged at the rim of the wheel. Figure 98 is a section of a turbine wheel. The buckets are represented in the outer portion, and the guides in the inner circle.

There are many kinds of turbines, and their effective power

Fig. 97.   Fig. 98.

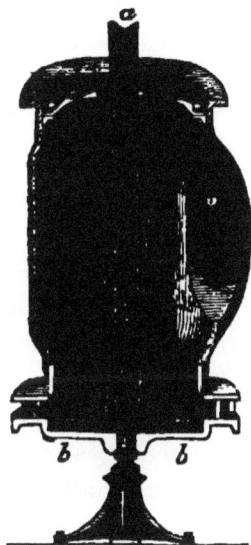

is from 75 to 88 per cent of that in the acting body of water. In the best form of overshot and breast wheels, it is from 65 to 75 per cent, and in undershot wheels from 25 to 33 per cent.

D. SOLIDS.

136. *Tendency of Solids to assume a Crystalline Structure.* — Solids, as a rule, *tend to assume a crystalline structure.* ¯This tendency is best shown by allowing a substance to pass gradually from a liquid to a solid state.

Place a rather dilute solution of acetate of lead (sugar of lead) in a tank with parallel sides of glass (such as is often used for projection), and fix two platinum wires in the solution, about an inch apart. Place the tank before the condenser of a magic lantern, and focus the wires on the screen. Connect the wires with the poles of a small voltaic battery. The lead will separate from the solution, and collect as a solid upon the wire connected with the negative pole of the battery. Beautiful fern-like forms will be seen to grow up on the screen. These forms are the *crystals of lead.* As the substance passes slowly from the liquid to the solid state, the molecules are free to arrange themselves according to their tendencies.

If alum is added to hot water as long as it will dissolve, and then the water is allowed to cool slowly, a part of the alum will be deposited on the bottom of the dish, — not in a confused mass, but in beautiful crystals. If saltpetre, nitrate of baryta, or corrosive sublimate is treated in the same way, beautiful crystals will be formed, but in each case the crystals will have a different shape.

Melt some sulphur in a crucible, and allow it to cool slowly

Fig. 99.

till a crust forms on the surface; then carefully break the crust and pour off the remaining liquid, and the crucible will be found lined with delicate needle-shaped crystals (Figure 99).

Large crystals of many solids can be obtained by dissolving as much of the solid as is possible in cold water, and then setting it away in a shallow dish where it will be free from dust and disturbance, and allowing the water to evaporate very slowly. *The more gradual the formation, the larger are the crystals.* The larger crystals seen in cabinets of minerals were probably centuries in forming. The water in which the solid was dissolved found its way into a cavity of a rock, and there slowly evaporated.

The tendency of the cohesive force to form the molecules into crystals is strikingly shown in cannon which have been many times fired, and in shafts of machinery and axles of car-wheels which are continually jarred. Such bodies often become brittle, and on breaking show the smooth faces of the crystals which have been formed. The continued jarring *gives the molecules a slight freedom of motion,* and crystals are slowly built up.

Many solids are crystalline in structure which do not appear to be so. Thus, a piece of ice is a mass of the most perfect crystals, but they are so closely packed together that we cannot readily distinguish them.

137. *Properties of Solids.* — A body is said to be *tenacious* when it is *difficult to pull it in two.* All solids are more or less tenacious, but they differ greatly in the degree of their tenacity. A body is said to be *hard* when it is *difficult to scratch or indent it,* that is to say, when it is *difficult to displace its molecules.* All solids are *elastic* within certain limits, and this elasticity may be developed by stretching, by bending, by twisting, and by compression, that is, by any kind of strain whatever. Different solids, however, differ greatly in the limit of their elasticity (9). When the strain is carried beyond the limit of elasticity, the body must either break or take up permanently a new form. A body which is *apt to break when strained beyond the limit of elasticity* is said to be *brittle.* A brittle substance is not always *easily* broken. Such a body will not break unless strained beyond the limit of its elasticity, and that is often a difficult thing to do. It is not easy to break a glass rod an inch in diameter, yet glass

is the most brittle substance known. Substances which can *readily take permanently new forms* are said to be *malleable* or *ductile.* A malleable substance is one that can be *hammered or rolled into sheets*, and a ductile substance one that can be *drawn into wire.* All malleable substances are to some extent ductile, but the most malleable are not the most ductile.

Gold is one of the most malleable of the metals. In the manufacture of *gold-leaf*, it is hammered out into sheets so thin that it takes from 300,000 to 350,000 of them to make the thickness of a single inch.

The gold is first rolled out into sheets by passing it many times between steel rollers in what is called a *rolling-machine.* The rollers are so arranged that they can be brought nearer to each other, pressing the gold into a thinner and thinner sheet every time it is passed between them. After it has thus been rolled out to the thickness of writing-paper, it is cut up into pieces about an inch square. These are piled into a stack with alternate pieces of tough paper, and beaten with wooden mallets. They are again cut up into small pieces, and arranged in a stack with alternate squares of gold-beater's skin, and again beaten with mallets. This last process is usually repeated three times.

II.

SOUND.

A. Origin of Sound.

138. *Sound originates in Molar Vibrations.* — Fix a point on a stand so as to be nearly in contact with a glass bell (Figure 100), and also hang a pith ball in contact

Fig 100.

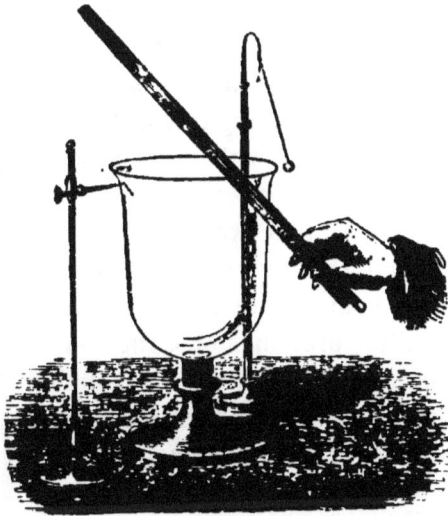

with the bell on the opposite side. If we draw a rosined bow across the edge of the bell, this will be made to emit a musical sound, and will also be heard to tap against the point, showing that it is *in vibration*. The pith ball will also be kept swinging as long as the sound continues. On touching the bell lightly, we feel that it is vibrating.

By grasping it firmly, we stop both the vibration and the sound.

Strike one prong of a tuning-fork, and hold it to the ear ; it is found to be emitting sound. Fill a glass brimful of water, and hold the edge of the prongs in contact with the water ; a shower of spray will fly off on each side, showing that the prong is *in vibration*.

When a string or wire is emitting a sound, it may often be *seen* to be vibrating. It assumes the form of an elongated spindle (Figure 101).

Fig. 101.

Fig. 102

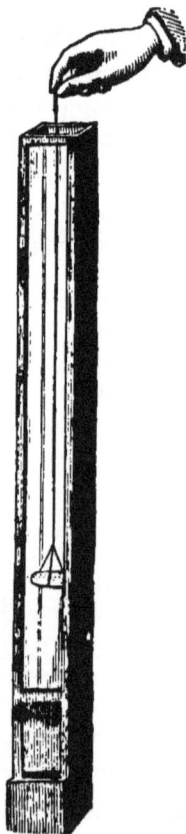

If the front of an organ pipe is made of glass, and a little stretched membrane covered with sand is lowered into it (Figure 102), when the pipe is emitting a sound, the sand will be seen to be agitated, showing that the air within the pipe is *in a state of vibration*.

By similar experiments it has been ascertained that every body which is emitting sound is in a state of molar vibration. When the vibration stops, the sound ceases. The more intense the vibration, the louder the sound. Sound, therefore, *originates in molar vibrations of ordinary matter*, solid, liquid, or gaseous.

139. *Fundamental and Harmonic Vibrations.* — Strew sand upon a horizontal plate of brass, and then, holding it with the thumb and finger (Figure 103), draw a bow across the edge of the plate so as to throw it into vibration. The sand will be tossed up and down at first, but will quickly come to rest in definite

lines, called *nodal lines*. These are *lines of rest* which separate the *vibrating segments* of the plate. By touching

Fig. 103.

the plates at different points with the thumb and fingers, a great variety of figures may be produced with the sand,

Fig. 104.

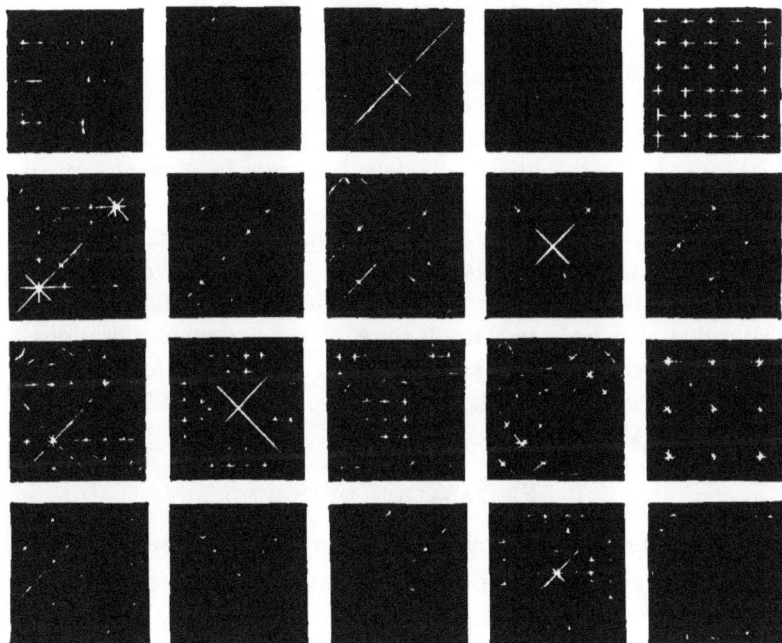

showing that it is possible for the plate to break up into vibrating segments in a great many different ways. A series of these nodal figures is shown in Figure 104.

Strings and columns of air may be also made to vibrate in segments. Figure 105 shows a string vibrating as a whole, in two segments, in three segments, and in four segments.

The vibration of a body as a whole is called its *fundamental* vibration; and the vibration of its segments, its *harmonic* vibration. The harmonic vibrations are more rapid than the funda-

Fig. 105.

mental vibrations. In a complete series of harmonic vibrations, the rate of vibration in the first harmonic is twice the fundamental rate; in the second harmonic, three times the fundamental rate : in the third harmonic, four times the fundamental rate; and so on.

It is not only possible to produce harmonic vibrations in a body, but it is *almost impossible not to produce them* when a body is thrown into vibration. Whenever the fundamental vibration of a body is started, some of the harmonic vibrations are almost certain to be started with it. Hence it follows that the molar vibrations of bodies which originate sound are *more or less complicated.*

B. Propagation of Sound.

140. *Sound is not propagated in a Vacuum.* — In Figure 106 the bell *B* is suspended by silk threads under the receiver of the air-pump. The bell is struck by means of

clock-work, which can be set in motion by the sliding-rod
r. If the bell is struck before exhausting the air, it can
be distinctly heard ; but as the air is exhausted, the sound
becomes fainter and fainter,
until at last it can hardly be
perceived, even with the ear
close to the receiver. Sound,
then, *cannot pass through a
vacuum.*

Fig. 106.

The slight sound which is
heard is transmitted by the
little air left in the receiver,
and by the cords which hold
up the bell.

141. *Sound is propagated in
Gases, Liquids, and Solids.* —
If hydrogen or any other gas
is now allowed to pass into
the receiver, the sound of the
bell is heard again. If a bell
is put under water and struck,
it can be heard. If a person
puts his ear close to the rail
of an iron fence, and the rail is struck at a considerable
distance, he hears the blow twice. The first sound comes
through the rail ; the second, which soon follows, comes
through the air. These experiments show that sound *passes
through gases, liquids, and solids.* Sounds are *propagated
chiefly by the air.*

142. *Sound is propagated by Waves.* — When any vibrating body, as the prong of a tuning-fork, is moving forward,
it crowds together the molecules of the air in front of it,
and so produces *a strain of compression* in the air. As the
body moves back again to its original position and beyond
it on the other side, it allows the molecules of the air

7

behind it to separate somewhat, and so produces *a strain
of rarefaction* in the air. Each of these strains is propagated through the air from molecule to molecule in precisely the same way that the strain of compression was
propagated from ball to ball in Figure 8. The molecules
of air in front of the vibrating body simply vibrate to and
fro with the sounding body. This vibrating motion is also
propagated from molecule to molecule through the air;
but while the strains of compression and rarefaction are
continually moving forward, each molecule of air moves
forward a short distance and then returns.

The *strains of compression and rarefaction* constitute what
is called a *sound-wave*, and each strain is called a *phase* of
the wave. If the body continues in vibration, the phases
of the waves will follow each other in regular succession.

The *distance occupied by the two strains* or phases is called
the *length* of the wave.

As the strain of compression is formed while the vibrating
surface is moving forward, and the strain of rarefaction while
the surface is moving backward, the length of each of these
phases will be the distance the strain propagates itself while the
sounding body performs half a vibration, and the length of
the sound-wave will be the distance the strain can propagate
itself while the sounding body is making a complete vibration.
Hence, the *faster the sounding body vibrates the shorter the
sound-waves, and the slower it vibrates the longer the waves.*

143. *The Intensity of Sound.* — The *intensity* of sound at
any point depends upon the *energy* of the vibration of the
molecules at that point.

As the sound-waves spread in all directions from the
sounding body, a greater and greater number of particles
of air must be set in motion, and the motion of each must
be more feeble; and since the surfaces of spheres increase
as the squares of their radii, the number of particles to be
set in motion increases as the square of the distance from

the sounding body. Sound, then, *diminishes in intensity as the square of the distance from the sounding body increases.*

If the sound-waves are prevented from spreading in all directions, the particles of air lose little of their motion, and the sound little of its intensity. Thus, Biot found that through one of the water-pipes of Paris words spoken in a very low tone could be heard at the distance of about three quarters of a mile. The sides of the pipe kept the sound-waves from spreading. In the same way conversation can be carried on between distant parts of a large building by means of small tubes, called *speaking-tubes.*

144. *The Velocity of Sound.* — The velocity of sound in air has been several times determined by experiment. In 1822 the French Board of Longitude chose two heights near Paris, and from the top of each fired a cannon at intervals of ten minutes during the night. The time between seeing the flash and hearing the report was carefully noted at both stations, and the average of the results showed that *sound travels through the air at the rate of* 1090 *feet a second.* In such experiments the time taken by the light to pass between the stations is too small to be perceived.

The velocity of sound in air depends somewhat upon the state of the atmosphere. Sound-waves travel faster with the wind than against it, and the higher the temperature of the air, the greater the velocity of sound in it. The velocity given above is for the temperature of 32°.

The velocity of sound in water is about 4700 feet a second, and its velocity in solids is still greater.

145. *The Reflection of Sound.* — When sound-waves meet the surface of a new medium, they are, in part, thrown back, or *reflected.* In this reflection, as in all cases of reflected motion, the angles of incidence and reflection are equal to each other.

Echoes are produced by the reflection of sound. In order to get an echo, we must have a reflecting surface far enough away to give an appreciable interval between the direct and reflected sounds. When the surface is less than 100 feet distant, the reflected sound blends with the direct sound.

The reflecting surface has often such a shape as to cause the different portions of the reflected wave to converge to a point, and so to intensify the reflected sound.

Multiple echoes may be produced by successive reflections from surfaces at different distances on the same side, or by alternate reflections from two surfaces on opposite sides. In some localities a pistol-shot is repeated thirty or forty times.

146. *The Speaking-Trumpet.* — The *speaking-trumpet* (Figure 107) consists of a long tube (sometimes six feet long), slightly tapering towards the speaker, furnished at this end with a hollow mouth-piece, which nearly fits the lips, and at the other with a funnel-shaped enlargement,

Fig. 107.

called the *bell*, opening out to a width of about a foot. It is much used at sea, and is found very effectual in making the voice heard at a distance. The explanation usually given of its action is, that the slightly conical form of the long tube produces a series of *reflections in directions more and more nearly parallel to the axis;* but this explanation fails to account for the utility of the *bell*, which experience has shown to be considerable.

147. *The Ear-Trumpet.* — The *ear-trumpet* is used by persons who are hard of hearing. It is essentially an *inverted speaking-trumpet*, and consists of a conical metallic tube, one of whose extremities, terminating in a *bell*, receives the sound, while the other end is introduced into the ear. This instrument is the reverse of the speaking-trumpet. The bell serves as a mouth-piece; that is, it receives the sound coming from the mouth of the person

who speaks. These sounds are transmitted by a series of reflections to the interior of the trumpet, so that the waves, which would become greatly developed, are *concentrated* on the auditory apparatus, and produce a far greater effect than divergent waves would have done.

148. *Loudness of Sound.* — The *loudness*, or *intensity*, of sound depends upon the *energy of the molecular vibrations* in the sound-waves. In a curve representing the form of the sound-wave, the loudness would be represented by the height of the curve, or the *amplitude* of the wave.

149. *Pitch of Sound.* — The *pitch* of sound depends upon the *rate at which the pulsations of sound strike upon the drum of the ear*, or upon the *length* of the sound-waves. The length of the sound-waves depends chiefly upon the *rate of vibration* of the sonorous body.

Two sounds are said to be in *unison* when the rate of vibration is *the same;* to form an *octave*, when their rates of vibration are as 2 to 1 ; a *fifth*, when their rates of vibration are as 3 to 2 ; a *fourth*, when their rates of vibration are as 4 to 3 ; and a *major third*, when their rates of vibration are as 5 to 4.

In the lowest note of the organ there are 16½ vibrations a second. In the lowest note of the piano there are 33 vibrations a second, and in the highest note 4224 ; giving a range of 7 octaves. In the highest note ever heard in an orchestra there are 4752 vibrations a second. This note is given by the piccolo flute. In the shrillest sounds that are audible there are about 32,000 vibrations a second, the upper limit of audibility varying with different persons. The voice of ordinary chorus-singers ranges from 100 to 1000 vibrations a second, and the extreme limits of the human voice are 50 and 1500 vibrations a second.

150. *Quality of Sound.* — The *quality* of sound depends upon the *form* of the sound-waves, that is, upon *the harmonic vibrations* which are present with the fundamental vibrations in the sonorous body. The *pitch* of sound is

determined chiefly by *the fundamental note*. Two sounds
of the same pitch may differ in quality, because of differ-
ences in their *harmonics*. *Fundamental* tones are those
produced by the fundamental vibrations of a sonorous
body ; and *harmonic* tones, those produced by the harmonic
vibrations. No two instruments or voices give tones of
the same quality, though they may be of the same loudness and pitch.

The difference between a *noise* and a *musical sound* is that
the latter is *smooth and regular*, and the former *rough and irregular*. Musical sounds are produced by rapid *periodic* vibrations
of a body, and noises by *non-periodic* vibrations.

151. *Interference of Sound.* — When two equal water-
waves meet *in the same phase*, namely, so that the crest of
one coincides with the crest of the other, and the hollow
of one with the hollow of the other, their combination
produces at the point of meeting a wave of *double the
height*. Were the two waves to meet *in opposite phases*,
that is, so that the hollow of one coincides with the crest
of the other, their combination would leave the surface of
the water undisturbed ; there would be *neither depression
nor elevation*.

In a similar way, when two equal sound-waves meet *in
the same phase*, their combination would produce at the
point of meeting a wave of *twice the degree of condensation
and rarefaction* of either of the component waves. Were
the two waves to meet *in opposite phases*, the air would be
undisturbed at the place of meeting; there would be
neither condensation nor rarefaction. An ear at the point
of meeting of the wave in the first case would hear a sound
much louder than that conveyed by either sound-wave
alone ; while in the second case it would hear *no sound at
all*. The *meeting of two sound-waves so as to neutralize each
other* is called the *interference* of sound.

Strike a tuning-fork so as to throw its prongs into vibration,

hold it vertically near the ear, and turn it slowly around so as to bring the sides, the edges, and the corners of the prongs successively towards the ear. Four positions of the fork will be found in which its sound will be in-   Fig. 108.
audible. Let *a* and *b* (Figure 108) be
the ends of the prongs of a tuning-fork
in vibration. The sound of the fork is
inaudible when the ear is on any one of
the dotted lines. As the prongs vibrate,
each develops a series of waves, and
along the dotted lines these two sets of
waves will be of equal intensity and in
opposite phases. Hence along these
lines the two sets of waves neutralize each other, and *silence results from the combination of two sounds.*

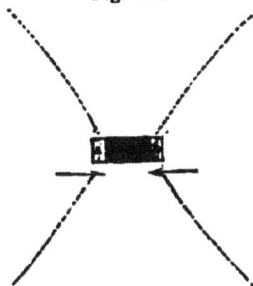

152. *Musical Beats.* — Suppose two tuning-forks, slightly different in pitch, to be started together, and suppose the prongs of both to be moving forward at the same time ; they will start waves of the same phase which will *coincide with and intensify each other.* The fork having the higher pitch will, however, immediately begin to *gain* on the other, and the coincidence of the waves will be less and less perfect until this fork has gained *half a vibration* on the other. The prongs of the two forks will now be moving in opposite directions at the same time, and the waves started by the two forks will be in opposition, and will *neutralize each other* wholly or in part. After this there will again be partial coincidence of the waves, and the degree of coincidence will increase till the higher fork has gained a *whole vibration* on the lower one, when the *coincidence will again be complete.* When two such forks are started together, the sound gradually dies away till it becomes nearly —or altogether inaudible ; it then swells out loudly, and gradually dies away again at regular intervals. These *gradual risings and fallings in the intensity of sound* are called *beats.*

These beats occur whenever two sounds of nearly the same pitch are produced together. The *rate* of beating will be *equal to the difference of the rate of vibration* in the two sonorous bodies. If one of the bodies gains one vibration a second on the other, the sounds will beat once a second ; if it gains two vibrations a second, the sounds will beat twice a second ; and so on.

Even after the beats become too rapid to be distinguished by the ear, they give a disagreeable roughness to the sound. According to Helmholtz, *dissonance* is entirely due to the roughness produced by *a rapid succession of beats*, which take place between either the fundamental tones or the harmonics which are present in the two sounds.

C. RESONANCE.

153. *Sympathetic Vibrations of Tuning-Forks.* — Take two tuning-forks of exactly the same pitch, cause one of them to vibrate, and hold it near the other without touching it. The second fork will soon begin to vibrate, and will emit a distinctly audible sound after the first has been stopped. The second fork will not be started by the first unless the two are of exactly the same pitch, as may be shown by sticking a little pellet of wax to the prong of one of the forks so as to diminish its rate of vibration. *Vibrations started in one body by the vibrations of another* are called *sympathetic* vibrations. The *production of sound by sympathetic vibrations* is called *resonance*.

The vibrations are communicated from one fork to the other by means of the air. The vibrations of the first fork produce condensations and rarefactions in the air which succeed each other at the rate at which the fork is vibrating. The number of condensations which would pass any point in a second is exactly equal to the number of vibrations executed by the fork in a second. In the condensations the pressure of the air is increased, and in the rarefactions it is diminished. Each condensation as it passes the prong of the second fork gives it a little push. As the second fork vibrates at exactly the same rate as the first, each condensation arrives in time to push the prong just as it is ready to move forward of itself; hence the prong is always pushed in the direction in which it is moving. The push of one condensation moves the prong but little, but the pushes are so timed that each moves it a little farther than the last, until the fork is made to vibrate strongly.

When the second fork cannot vibrate at the same rate as the first, the condensation will sometimes push in the direction in which the prong is moving and sometimes in the opposite direction. Hence one push will neutralize the effect of another instead of augmenting it.

154. *Sympathetic Vibrations of Strings.* — If a piano is opened and one of the keys gently depressed so as to raise the damper without striking the string with the hammer, and the note of the string is then sung over the piano, the string will begin to vibrate and will emit an audible sound for a little time after the voice ceases. It is only necessary to hit the pitch of a string accurately and to sustain the note sufficiently. Strings may be thrown into vibration by their harmonic notes as well as by their fundamental notes.

155. *Sympathetic Vibrations of Masses of Air.* — If a vibrating tuning-fork is held at the end of a tube an inch and a half or two inches in diameter, the sound of the fork will be powerfully reinforced, provided the tube is of suitable length. The suitable length for a tube open at both ends is one half of the length of the wave produced by the fork. A tube closed at one end resounds most powerfully when its length is one quarter of the length of the wave produced by the fork. The column of air in the tube is thrown into powerful sympathetic vibrations by the fork, and these vibrations greatly augment the sound. The moment the fork is stopped the resonance ceases.

Columns of air may also be thrown into sympathetic vibration by their harmonic vibrations. By altering the shape of the tube it may be made to reinforce certain harmonics more powerfully than others, and so change the quality of the exciting sound.

156. *Sounding Boards and Boxes.* — The sound of a tuning-fork is feeble unless reinforced by *a resonant case* of suitable dimensions to which the fork is fixed. Such a resonant case is called a *sounding-box.*

Thin pieces of dry straight-grained pine, such as are employed for the faces of violins and the sounding-boards of pianos, are capable of vibrating more or less freely, in any

period lying between certain wide limits. They are accordingly
set in vibration by all the notes of their respective instruments;
and by the large surface with which they act upon the air, they
contribute in a very high degree to increase the sonorous effect.
All stringed instruments are provided with sounding-boards;
and their quality mainly depends on the greater or less readiness with which these respond to the vibrations of the strings.

D. MUSICAL INSTRUMENTS.

157. *Stringed Instruments.* — In one class of musical
instruments the notes are produced by *the transverse vibrations of strings.* These instruments are called *stringed*
instruments. The *rate* at which a string vibrates depends
upon its *length,* its *weight,* and its *tension.* The *shorter,*

Fig. 109.

the *tighter,* and the *lighter* a string, the *faster* it vibrates.
Strings may be thrown into transverse vibration by draw-
ing a rosined bow across them, as in the case of the violin;
or by plucking them with the finger, as in the case of the
harp; or by striking them with a hammer, as in the case
of the piano.

In the piano there is a string for every note. In the violin
and similar instruments, several notes are obtained from the
same string by fingering it so as to change its length and
tension.

158. *The Sonometer.* — The *sonometer* (Figure 109) is an
instrument for *investigating the laws of the vibration of
strings.* It consists essentially of a string or wire stretched
over a sounding-box by means of a weight. One end of

the string is secured to a fixed point at one end of the sounding-box ; the other end passes over a pulley, and carries weights which can be altered at pleasure. Near the two ends of the box are two fixed bridges, over which the cord passes. There is also a movable bridge, which can be employed for altering the length of the vibrating portion.

159. *Wind Instruments.* — In *wind* instruments the notes are produced by *the longitudinal vibrations of columns of air enclosed in pipes.* The *rate* of vibration depends upon the *length* of the column, and upon *whether the pipe is opened or closed.* The *shorter* a column of air the *faster* it vibrates, and the air in an *open* tube vibrates *twice as fast* as that in a *closed* pipe of the same length. This is because the air in a closed pipe vibrates as a whole, while that in an open pipe vibrates in two segments, there being a stationary point or *node* at the centre of the pipe.

In an organ there are as many pipes as notes, only one note being obtained from each pipe. In the case of the flute and similar wind instruments, several notes are obtained from one pipe by opening and closing the holes at the side of the pipe so as to alter the length of the vibrating column of air, and by altering the strength of the blast so as to change from the fundamental note of the pipe to one or other of its harmonics.

In all wind instruments the pipe is made to speak by *resonance.* The sympathetic vibrations in the pipe are sometimes started by the vibrations of the lips, as in the case of the trumpet ; or by the vibrations of a spring called a *reed*, as in the case of the clarionet ; or by the flutter of a jet of air when blown against a sharp edge, as in the case of the flute.

160. *Organ Pipes.* — Organ pipes are made of wood or metal, and they are made to sound either by blowing *against a sharp edge so as to produce a flutter*, or by blowing *against a spring so as to throw it into vibration.* Pipes which are made to sound in the first way are called *flue-pipes ;* and those made to sound in the second way, *reed-pipes.* Pipes *closed at one end* are called *stopped* pipes ; and those *open at both ends* are called *open* pipes.

Two forms of flue-pipes are shown in Figures 110 and 111 ; the one being made of wood, the other of metal. The air passes from the bellows through the tube *P* into a chamber, which is closed at the top except the narrow slit *i*. The air compressed in the chamber passes through this slit in a thin sheet, which breaks against a sharp edge *a*, and there produces a flutter. The space between the edge *a* and the slit below is called the *mouth* of the pipe. The metal *reed* commonly used in organ

Fig. 110.   Fig. 111.

pipes is shown in Figures 112 and 113. It consists of a long strip of flexible metal *V V*, placed in a rectangular opening, through which the current of air enters the pipe. As soon as the air begins to enter the pipe, the force of the blast bends down the spring of the reed so as to close the opening. The elasticity of the reed causes it to fly back at once, so as to open the pipe and allow the air to enter again. It thus breaks up the current of air into a regular succession of little puffs.

161. *The Organ of the Human Voice.* — The organ of voice in man is situated at the top of the windpipe, or *trachea*, which is the tube through which the air is blown from the lungs. A pair of

Figs. 112, 113.

elastic bands, called the *vocal chords*, stretched across the top of the windpipe so as nearly to close it, form *a double reed.* When air is forced from the lungs through the slit between the chords, these are made to vibrate. By changes in their tension, their rate of vibration is varied, and the sound raised or lowered in pitch. The cavity of the mouth and nose acts as *a resonant tube*, and by altering the shape of this cav-ity we can give greater promi-nence to either the fundamental note of the vocal chords or to any of their harmonics.

162. *Singing Flames.* — The air in an open tube may be made to give a sound by means of a luminous jet of hydrogen, coal gas, etc. When a glass tube about twelve inches long is held over a lighted jet of hydrogen (Figure 114), a note is produced which, if the tube is in a cer-tain position, is the fundamental note of the tube. The current of air passing up through the

Fig. 114.

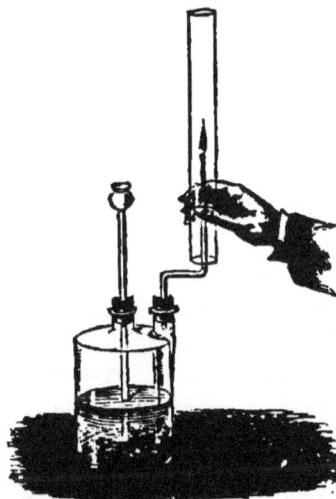

tube over the flame causes the flame to flutter, and *the air in the tube reinforces some pulsations of this flutter by sympathetic vibration.* The vibration of the column of air in the tube re-acts upon the flame, and causes it to vibrate more regularly and

more powerfully. The note depends on the size of the flame
and the length of the tube.

If, while the tube emits a certain sound, the voice is gradually raised to the same pitch, as ·soon as the note is nearly
in unison with that of the tube, the flame is agitated, jump-
ing up and down, but becomes steady when the two sounds
are in unison. If the note is then gradually raised in pitch,
the pulsations again commence; they are the optical expres-
sions of the *beats* which occur near perfect unison. If, while
the jet burns in the tube, and produces a note, the position
of the tube is slightly altered, a point is reached at which no
sound is heard. If now the voice or the tuning-fork is pitched
at the note produced by the jet, it begins to sing, and continues
to sing even after the voice or fork is silent. A mere noise or
shouting at an incorrect pitch affects the flame, but does not
cause it to sing.

Fig. 115.

163. *Edison's Phonograph.* — In Edison's *phonograph*, the
vibrations of the air are first taken up by a thin plate of metal,
and are then permanently registered on a sheet of tin-foil.
This instrument (Figures 115 and 116) consists essentially
of a brass cylinder C and of a mouth-piece F. On the surface
of the cylinder is constructed a very accurate spiral groove, the
threads of which are about $\frac{1}{10}$ of an inch apart. The cylinder is
turned by the crank D upon the axis $A B$. On one end of this
axis is cut a thread of the same fineness as the groove on the
cylinder. A sheet of tin-foil is fastened smoothly on the surface
of the cylinder. The mouth-piece (Figure 116) is supported on
a post G, and may be moved to and from the cylinder by the
lever H. At the bottom of the mouth-piece there is an iron
plate A about $\frac{1}{100}$ of an inch thick. Under this plate are two

pieces of rubber tubing x and x, which separate it from a spring supported by E, and carrying a round steel point P, which rests upon the tin-foil on the cylinder, just over the spiral groove. If the crank is turned, the thread on the axis causes the cylinder to move forward so as to keep the groove always under the point. When the iron plate is at rest, if we turn the crank the point marks a spiral line of uniform depth on the tin-foil. If we speak or sing into the mouth-piece, the vibrations of the air are communicated to the iron plate, and from this to the point

Fig. 116.

by means of the rubber tubing. If the crank is turned while a person is speaking or singing into the mouth-piece, the point will mark a dotted line on the tin-foil. The depth of the indentations will *exactly represent the densities of the different portions of the sound-waves* which encounter the disc. The forms of the sound-waves are thus registered on the tin-foil, and may be studied at leisure with the microscope.

If, after talking into the mouth-piece, we set the cylinder back to the starting-point and then turn the crank, the point will follow the indentations in the tin-foil, and so be compelled *to vibrate exactly as it did when it made these indentations* in the foil. The vibrations of the point will be communicated to the thin iron plate by means of the rubber, and by the plate to the air. Thus the words spoken into the mouth-piece will be exactly repeated, and by the use of a properly constructed mouth-piece they may be rendered audible throughout a large hall. By resetting the cylinder they may be repeated several times, though more feebly each time the foil is passed under the point, the indentations being gradually smoothed out.

E. The Human Ear.

164. *The Human Ear.* — A section of the ear is shown in Figure 117. The external opening is closed at the bottom by a circular membrane called the *tympanum*, behind which is the cavity called the *drum* of the ear. This cavity is separated from the space between it and the brain by a bony partition, in which are two openings, the one round and the other oval. These also are closed by delicate membranes. Across the cavity of the drum stretches a series of four little bones : the first, called the *hammer*, is attached to the tympanum ; the second, called the *anvil*, is connected by a joint with the hammer ; a third little round bone connects the anvil with the *stirrup* bone, which has its oval base planted against the membrane of the oval opening, almost covering it. Behind the bony partition, and between it and the brain, is the *labyrinth*, which is filled with water, and over the lining of which the fibres of the *auditory nerve* are distributed.

The tympanum intercepts the vibrations of the air in the external ear, and transmits them through the series of bones in the drum to the membrane which separates the drum from the labyrinth ; and thence to the liquid within the labyrinth itself, which in turn transmits them to the nerves. The transmission, however, is not direct. At a certain place within the labyrinth, exceedingly fine elastic bristles, terminating in sharp points, grow up between the nerve fibres. These *bristles of Schultze* (so called from the discoverer) are exactly fitted to sympathize with those vibrations of the water which correspond to their proper periods. Thrown thus into vibration, the bristles stir the nerve fibres which lie between their roots, and the nerve transmits the impression to the brain, and thus to the mind. At another place in the labyrinth we have little crystalline particles, calléd *oto-liths*, embedded among the nervous filaments, and exerting, when they vibrate, an intermittent pressure upon the adjacent nerve fibres. The otoliths appear to be fitted, by their weight, to receive and prolong the vibrations of evanescent sounds which might otherwise escape attention. The bristles of Schultze, on the contrary, are peculiarly fitted for the transmission of continuous vibrations. Finally, there is in the labyrinth the *organ of Corti* (named from the discoverer), which is to all appearance

Fig. 117.

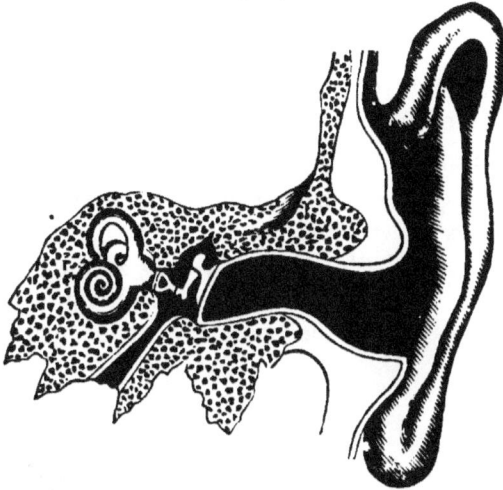

a musical instrument, with its chords so stretched as to receive vibrations of different periods, and transmit them to the nerve filaments which traverse the organ. Within the ear of man, and without his knowledge or contrivance, this lute of 3000 strings has existed for ages, receiving the music of the outer world, and rendering it fit for reception by the brain. Each musical tremor which falls upon this organ selects from its tense fibres the one appropriate to its own pitch, and throws that fibre into sympathetic vibration. And thus, no matter how complicated the motion of the external air may be, these microscopic strings can analyze it, and reveal the elements of which it is composed. ·

8

III.

HEAT.

I. EFFECTS OF HEAT.

A. EXPANSION.

165. *Expansion of Solids.* — As a rule, *bodies expand when heated*, solids being the least expansible, liquids next, and gases the most expansible.

The *linear expansion* of a solid may be illustrated by means of the apparatus shown in Figure 118. The metal rod *A* is supported on two standards. It is fastened at the end *B* by the binding screw; the other end passes loosely through its standard, and presses against the short arm of the index *K*, which

Fig. 118.

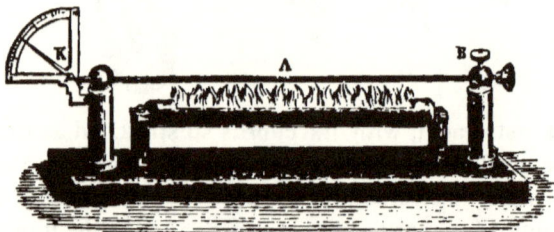

moves over a graduated arc. Under the rod there is a vessel filled with alcohol. The rod is adjusted so that the index shall be at zero on the scale, and the alcohol is lighted. As the rod becomes heated, the index rises, showing that the rod has expanded in length so as to move forward the short arm of the index.

If a brass and iron rod of the same length and thickness are tried in succession, and each is raised to a bright red heat, it will be found that the brass rod will expand considerably more than the iron. As a rule *different solids expand unequally when heated equally.*

The *cubical expansion* of a solid may be illustrated by means of the ring and ball shown in Figure 119. When cool, the ball will just pass through the ring. If we heat the ball by holding it for a time in the flame of the lamp, it will no longer pass through the ring ; but if allowed to cool, it will again pass through. If, while the heated ball rests on the ring, this is heated equally with the ball, the latter will again pass through the ring, the two being equally expanded by the heat.

166. *Force of Expansion of Solids.* — The *force* of expansion is very great, being *equal to that which would be*

Fig. 119.

necessary to compress the body to its original dimensions. Thus, for instance, iron when heated from 32° to 212° increases by .0012 of its original length. In order to produce a corresponding change of length in a rod an inch square, a force of about 15 tons would be required.

It would be useless to attempt to offer any mechanical resistance to a force so enormous ; the only thing that can be done, in the case of structures in which metals are employed, is to arrange the parts in such a manner that the expansion shall not be attended with any evil effects. Thus, in a railway, the rails do not touch each other, a small interval being left to allow room for the variations of length. Iron beams employed in buildings must have the ends free to move forward, without encountering any obstacles, which they would inevitably overthrow. Sheets

of zinc or lead employed in roofing are so arranged as to be able
to overlap one another on expansion.

167. *Compensating Pendulum.* — Suppose a clock to keep
exact time at a certain temperature ; then, if the temperature
rises, the length of the pendulum will increase (62), and with it
the duration of each oscillation, so that the clock will lose.
The opposite effect would be produced by a fall of temperature.
Hence the clock is liable to go too fast in winter, and too slow in

Fig. 120.   Fig. 121.   Fig. 122.

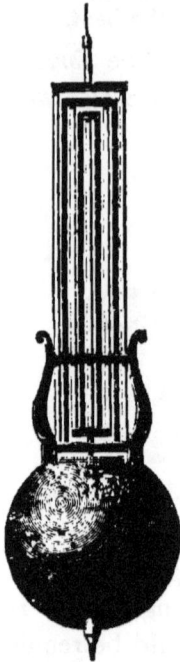
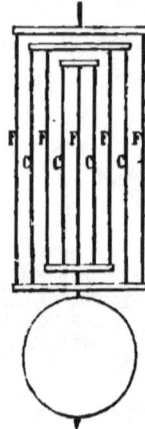

summer ; and we must move the ball of the pendulum from time
to time in order to insure its regularity.

The effect of temperature may be notably diminished by
means of *compensating pendulums,* of which there are several
different kinds.

Harrison's gridiron pendulum (Figures 120 and 121) consists
of four oblong frames, the uprights of which are alternately of

brass, C, and of steel, F. These are so put together that the expansion of the steel rods alone would tend to lower the ball, while the expansion of the brass rods alone would tend to raise it. The lengths of the rods are so adjusted that the expansion of one set of rods shall just balance that of the other, thus keeping the ball of the pendulum all the time at exactly the same distance from the point of suspension.

Graham's pendulum consists of an iron rod carrying at the bottom a frame which holds one or two tubes containing mercury (Figure 122). The mercury takes the place of the ball of the pendulum. The expansion of the rod tends to lower the centre of gravity of the mercury, while the expansion of the mercury, since it is free to expand only upward, tends to raise the centre of gravity. The quantity of mercury is adjusted so that its expansion shall balance that of the rod, and thus keep the centre of gravity of the mercury at the same height all the time.

Fig. 123.

168. *Compensation Balance-Wheel.* — The rate of a watch is controlled by the vibration of the *balance-wheel*. The larger this wheel the slower it vibrates, and the smaller it is the faster it vibrates. Hence changes of temperature have the same effect on the rate of watches as on that of clocks. The rim of the *compensation balance-wheel* (Figure 123) is made in sections, which are weighted at their free ends, and are composed of two metals, the more expansible of which is on the outer side of the sections. The expansion of the spokes tends to carry the weights away from the centre of the wheel and so to make the wheel larger. When the sections of the rim expand, they become more curved, since they expand more rapidly on the outside than on the inside; hence they tend to carry the weight in towards the centre and so to make the wheel smaller. The parts of the wheel are so adjusted that the expansion of the sections of the rim just balances that of the spokes.

169. *Expansion of Liquids.* — The expansion of a liquid may be illustrated by means of a bulb with a projecting

Fig. 124.

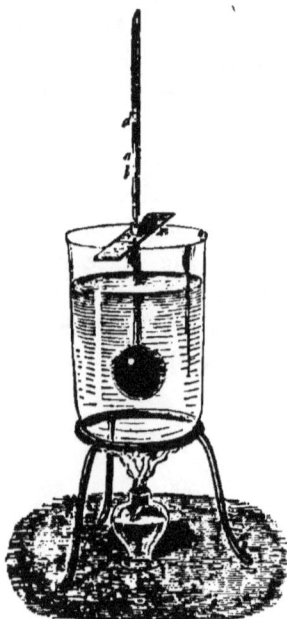

tube (Figure 124), filled with water or other liquid up to the point *a*. If the bulb is immersed in a vessel of hot water, the liquid in the stem at first falls to *b*, and then gradually rises to *a*. The liquid falls at first, because the bulb, being the first heated, is also the first to expand, and its capacity is thus increased. Afterwards, as the liquid becomes heated, it expands more rapidly than the globe; hence it rises in the tube.

If two bulbs, with projecting tubes, and of exactly the same size, are filled, one with water and the other with alcohol, and are then heated equally, the alcohol will be seen to expand more rapidly than the water. In general, *different liquids when heated equally expand unequally.*

170. *Anomalous Expansion and Contraction of Water.* — If we fill a bulb and tube with water, and surround the bulb with a freezing mixture, the water in the stem will steadily fall till the temperature of the water has reached 39°; it will then begin to rise again, and will continue to rise till the temperature reaches 32°. If now the bulb is gradually heated. the water will fall in the stem till the temperature reaches 39°; it will then begin to expand, and will continue to expand until it boils. *Water at 39° will expand whether it is heated or cooled.* It follows from this, that water is at its greatest density at 39°. Hence this point of temperature is called its *point of maximum density.*

171. *Expansion of Gases.* — The expansion of air may

be illustrated by means of the bulb and
tube shown in Figure 125. The bulb is
filled with air, which is separated from
the external air by a small column of
liquid in the stem, which serves also as
an index. When the globe is warmed
by the hands, the index is rapidly pushed
up. It has been found that *all gases
expand equally for the same rise of temperature*, and that under a uniform pressure a gas will expand so as to double
its volume for a rise of temperature of
about 490°.

Fig. 125.

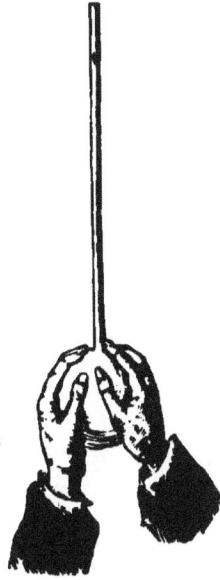

172. *Expansion due to an Increase of
Molecular Motion.* — The molecules of bodies are all the time moving rapidly to and
fro. When heat is applied to a body, its
molecules are made to move more rapidly,
and this increased agitation causes them to move farther apart,
and the body to expand.

B. Measurement of Temperature.

173. *Temperature.* — When we wish to indicate *how hot*
a body is, we say that it has a certain *temperature*. The
word *temperature* is the noun which corresponds to the
adjective *hot*. We estimate how hot a body is from its
power of imparting heat to other bodies. The body which
has the greater power of imparting heat is said to be the
hotter, or to have the higher temperature.

*Temperature is the thermal condition of a body considered
with reference to its power of imparting heat to other bodies.*

An instrument used *for measuring temperature* is called a
thermometer.

174. *The Mercurial Thermometer.* — In ordinary thermometers changes of temperature are indicated and meas-

Fig. 126.

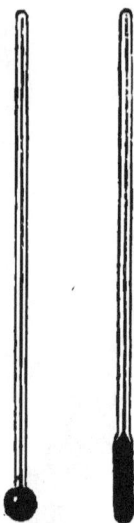

ured by *the expansion and contraction of mercury.*
The instrument is called a *mercurial thermometer.* It consists essentially (Figure 126) of a
tube with a very fine calibre, closed at one
end, and having a reservoir at the other end,
usually in the form of a globe or cylinder.
The bulb and a portion of the stem are filled
with mercury. As the temperature changes,
the top of the column of mercury in the tube
rises and falls. A scale is either engraved on
the stem or placed behind it.

175. *How the Mercurial Thermometer is
Graduated.* — The two *fixed points of temperature* are those at which *ice melts* and *water boils.*
The former is called the *freezing-point*, and the
latter the *boiling-point.*

In order to determine the position of the freezing-point on the
stem, the bulb and the lower part of the stem are surrounded by

Fig. 127.

melting ice, contained in a perforated
vessel so as to allow the water produced by the melting to escape (Figure
127). When the column in the stem
ceases to fall, a mark is made on the
tube, with a fine diamond, at the top of
the mercurial column. This mark indicates the position of the freezing-
point for this particular thermometer.

In order to obtain the position of
the boiling-point, the bulb and stem
of the thermometer are enveloped in
steam from boiling water, as shown in
Figure 128. The height to which the mercury rises is then
marked on the stem.

176. *The Fahrenheit and Centigrade Scales.* — There are
two thermometer *scales* in common use, the *Fahrenheit* and
the *Centigrade.* The ordinary scale in use in this country

and in England is the Fahrenheit scale. On this scale *the freezing-point is marked* 32 *and the boiling-point* 212. The space between the freezing and boiling points is divided into 180 equal parts, each of which is called a *degree.* These divisions are continued on the scale above the boiling-point and below the freezing-point to the ends of

Fig. 128.

the tube. A Fahrenheit degree is $\frac{1}{180}$ of the difference of temperature between the freezing and boiling points.

On the *Centigrade* scale *the freezing-point is marked* o *and the boiling-point* 100, and the space between the two is divided into 100 equal parts, the divisions being continued to the ends of the tube. A Centigrade degree is $\frac{1}{100}$ of the difference of temperature between the freezing and boiling points. *A Fahrenheit degree is ⅝ of a Centigrade degree.*

The *zero* of the *Centigrade* scale is *the temperature of melting ice.* The *zero* of the *Fahrenheit* scale is 32° *below the melting-point of ice.* It was the lowest temperature that Fahrenheit could obtain with a mixture of salt and ice.

177. *Alcohol Thermometers.* — Mercury freezes at a temperature of about 40° below zero, or of — 40° F.; hence it cannot be used for measuring temperatures below that point. Low temperatures are sometimes measured by means of an *alcohol thermometer.* This is constructed in the same way as a mercurial thermometer, but the bulb is filled with alcohol instead of mercury. As alcohol boils at a temperature of about 175° F. an alcohol thermometer cannot be used for measuring high temperatures.

178. *Pyrometers.* — Mercury boils at a temperature of about 670° F. ; hence it cannot be employed to measure temperatures above that point. Very high temperatures are often measured by *the expansion of solids.* The instrument used is called a *pyrometer.* One form of pyrometer (Figure 129) consists of a

Fig. 129.

bar of iron lying in the groove of a porcelain slab. One end of the iron bar presses against the end of the groove, and the other end against the arm of an indicator. As the bar expands it moves the index point, the position of which indicates roughly the temperature to which the bar is exposed. Such pyrometers are not very accurate.

179. *The Differential Thermometer.* — Leslie's *differential thermometer* (Figure 130) enables us to measure *small variations of temperature.* A column of sulphuric acid, colored red, stands in the two branches of a bent tube, which terminates in two globes of equal volume. When the air contained in

the two globes is at the same temperature, the liquid stands
at the same height in the two
branches. This point is marked
zero. One of the globes being then
maintained at a constant temperature, the other is raised through,
for instance, 5 degrees, when the
column rises on the side of the
colder globe up to a point *a*, and
descends on the other side to a
point *b*. Suppose the space traversed by the liquid in each branch
to be divided into 10 equal parts,
each part will be equivalent to a
quarter of a degree. This division
is continued upon each branch on
both sides of zero.

Fig. 130.

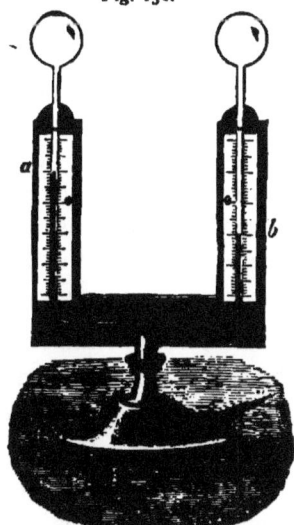

QUESTIONS ON THERMOMETER SCALES.

72. Oil of vitriol freezes at − 30° F. This is equivalent to
what temperature on the Centigrade scale?

73. Lead melts at 620° F. What is the temperature at
which lead melts on the Centigrade scale?

74. Iron melts at 2800° F. What is the equivalent temperature on the Centigrade scale?

75. What temperature on the Fahrenheit scale corresponds
to 50° C.? To − 25° C.? To 380° C.?

C. CHANGE OF STATE.

I. FUSION AND SOLIDIFICATION.

180. *The Fusing-Point.* — When any solid is sufficiently
heated it will melt, but *different solids melt at very different
temperatures.* The temperature at which a solid melts is
called its *melting-point* or *fusing-point.* Mercury melts at
− 40° F., ice at 32° F., lead at 608° F., and silver at 1832° F.

Most substances expand on melting, but a few, like ice, con-
tract. When a substance expands on melting, an increase of

pressure upon it will tend to hinder its melting, and will therefore raise its melting-point; but if it contracts on melting, an
increase of pressure will tend to help its melting, and will
accordingly lower its melting-point.

The passage from the solid to the liquid state is generally
abrupt, but this is not always the case. Glass, for instance,
before reaching a state of perfect liquefaction, passes through
a series of intermediate stages in which it is of a viscous con-
sistency, and can be easily drawn out into exceedingly fine
threads, or moulded into different shapes.

181. *Constant Temperature during Fusion.* — During the
entire time of fusion *the temperature remains constant.*
Thus, if a vessel containing ice is placed on the fire, the
ice will melt more quickly as the fire is hotter ; but if the
mixture of ice and water is constantly stirred, a thermometer placed in it will indicate the temperature 32° without
variation, so long as any ice remains unmelted ; it is only
after all the ice has become liquid that a rise of temperature will be observed.

182. *Latent Heat of Fusion.* — As we have just seen, *all
the heat that enters the body while it is undergoing fusion is
employed in changing its state.* The heat thus employed is
said to be rendered *latent*, and is called the *latent heat of
fusion*, or, since it exists in the latent state in the liquid
formed, the *latent heat of the liquid.*

183. *Solidification.* — Were any substance sufficiently
cooled, it would become solid. This *conversion of a substance into a solid by a reduction of temperature* is called
solidification, or *congelation.*

184. *Change of Volume in Congelation.*—In passing from
the liquid to the solid state, *bodies generally undergo a
diminution of volume ;* there are, however, exceptions, such
as ice, bismuth, silver, cast-iron, and type-metal. It is this
property which renders these latter substances so well
adapted for *casting*, as it enables the metal to penetrate
completely into every part of the mould.

The expansion of *ice* is considerable, amounting to about $\frac{1}{14}$ of its bulk; its production is attended by *enormous mechanical force*, just as in the analogous case of expansion by heat (166). Its effect in bursting water-pipes is well known. Major Williams at Quebec filled a 12-inch shell with water, and closed it with a wooden plug, driven in with a mallet. The shell was then exposed to the air, the temperature being — 18° F. The water froze, and the plug was projected to a distance of more than 100 yards, while a cylinder of ice of about 8 inches in length was protruded from the hole. In another experiment the shell split in halves, and a sheet of ice issued from the rent (Figure 131).

Fig. 131.

It is the expansion and consequent lightness of ice which enables it to float on the surface of the water, and to protect animal life beneath.

II. EVAPORATION AND CONDENSATION.

185. *Evaporation of Liquids.* — The majority of liquids, when left to themselves in contact with the atmosphere, *evaporate*, that is, gradually *pass into the state of vapor and disappear*. This occurs much more rapidly with some liquids than with others, and those which evaporate most readily are said to be the most *volatile*. Thus, if a drop of ether is let fall upon any substance, it disappears almost instantaneously; alcohol also evaporates very quickly, but water requires a much longer time. The change is in all

cases *accelerated by an increase of temperature;* in fact, when we *dry* a body before the fire, we are simply availing ourselves of this property of heat to hasten the evaporation of the moisture of the body. Evaporation may also take place from solids.

186. *Gas and Vapor.* — The words *gas* and *vapor* have no essential difference of meaning. A *vapor* is *the gas into which a liquid is changed by evaporation.* Every gas is probably the vapor of a certain liquid. The word *vapor* is especially applied to *the gaseous condition of bodies, which are usually met with in the liquid or solid state*, as water, sulphur, etc. ; while the word *gas* generally denotes a body which, *under ordinary conditions, is never found in any state but the gaseous.* When the air or any other gas *contains all the vapor it can hold*, it is said to be *saturated* with that vapor. The amount of vapor required to saturate a gas increases with the temperature. This may be shown by the following experiment. Pour a few drops of water into a glass flask, and then apply heat till the water is entirely evaporated and the flask appears dry. If the flask is allowed to cool, moisture will collect on its inner surface.

187. *Dry Air and Currents of Air favorable to Evaporation.* — The dryer the air the more rapid the evaporation, because the more readily will the atmosphere take up the vapor formed. Currents of air favor evaporation, because they prevent any layer of air from remaining long enough in contact with the liquid to become saturated with vapor. Other things being equal, wet clothes will dry much faster on a windy day than on a still day.

188. *Latent Heat of Evaporation.* — Evaporation is a *cooling process.* If a few drops of ether are allowed to fall on the hand, they will evaporate rapidly, and a sensation of cold will be experienced. If the bulb of a thermometer is dipped in ether and removed, the ether which adheres

to it will quickly evaporate, and the mercury will fall several degrees. The *heat consumed in evaporating a liquid* is called the *latent heat of evaporation*, or the *latent heat of the vapor*.

189. *Ebullition.* — When a liquid contained in an open vessel is subjected to a continual increase of temperature, it is gradually changed into vapor. This action is at first confined to the surface; but after a certain time bubbles of vapor are formed in the interior of the liquid, which rise to the top, and set the entire mass in motion with a characteristic noise; this is what is meant by *ebullition*, or *boiling*.

If we observe the gradual progress of this phenomenon, — for example, in a glass vessel containing water, — we shall perceive that after a certain time very minute bubbles are given off; these are bubbles of dissolved air. Soon after, at the bottom of the vessel, larger bubbles of vapor are formed, which decrease in volume as they ascend, and disappear before reaching the surface. This stage is accompanied by a peculiar sound, and the liquid is said to be *singing*. The sound is probably caused by the collapsing of the bubbles as they are condensed by the colder water through which they pass. Finally, the bubbles increase in number, growing larger as they ascend, until they burst at the surface, which is thus kept in a state of agitation; the liquid is then said to *boil*.

190. *Difference between Evaporation at the Boiling-Point and below the Boiling-Point.* — Below the boiling-point evaporation takes place only at the surface; the *tension*, or elastic force, of the vapor is less than that of the atmosphere; and only a part of the heat received by the liquid is used in converting the liquid into vapor, the temperature of the liquid rising all the time that heat is applied to it. At the boiling-point evaporation takes place throughout the liquid; the tension of the vapor formed is equal to that of the atmosphere; and *all the heat received by the liquid is used in converting it into steam*, the temperature

remaining stationary. The elastic force of the vapor given off by a liquid increases with the temperature, until· we reach the boiling-point, when it equals that of the atmosphere. The boiling-point of a liquid is therefore *the temperature at which the elasticity of the vapor is equal to the pressure of the atmosphere on the surface.* It follows from this that the boiling-point *must vary with the pressure.* Under a pressure less than that of the atmosphere the boiling-point of water is below 212°, and under a greater pressure than that of the atmosphere is above 212°.

191. *Franklin's Experiment.* — Boil a little water in a flask long enough to expel all the air from the flask. Remove the

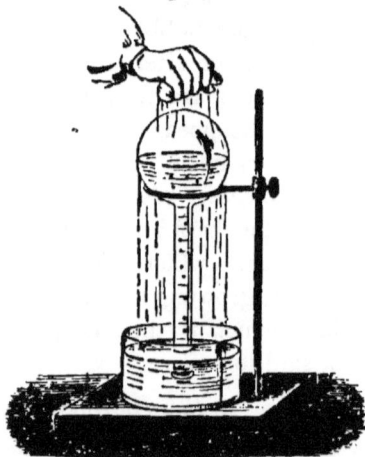

Fig. 132.

flask from the source of heat, cork it securely, and invert it with its corked end under water. Ebullition ceases almost instantly. Pour cold water over the flask (Figure 132) and the liquid will begin to boil, and will continue to do so for some time. The contact of the cold water with the flask lowers the temperature and tension of the steam which presses on the surface of the water, and *the diminution of pressure allows the water to boil at a lower temperature.*

192. *Papin's Digester.* — In a confined vessel water may be *raised to a higher temperature than in the open air,* but it *will not boil.* This is the case in the apparatus called, from its inventor, *Papin's digester* (Figure 133). It is a bronze vessel of great strength, covered with a lid secured by a powerful screw. It is employed for raising water to very high temperatures, and thus obtaining effects which would not be possible with water at 212°, such, for example, as dissolving the gelatine contained in bones.

*The tension of the steam increases rapidly with the tempera-
ture.*. Thus, at 392° the pressure is that of 16 atmospheres, or
about 240 pounds 'on the square inch. In order to obviate the
risk of explosion, Papin introduced the device known as the
safety-valve. It consists of
an opening, closed by a conical valve or stopper, which is
kept down by a lever loaded
with a weight. Suppose the
area of the lower end of the
stopper to be 1 square inch,
and that the pressure is not to
exceed 10 atmospheres, corresponding to a temperature
of 356°. The magnitude and
position of the weight are so
arranged that the pressure
on the whole is 10 times 15
pounds. If the tension of
the steam exceeds 10 atmos-
pheres, the lever will be raised, the steam will escape, and the
pressure will be thus relieved.

Fig. 133.

193. *Condensation of Vapors.* — *Condensation,* or the
conversion of a vapor into a liquid, is the reverse of evaporation. In condensation, the heat rendered latent in
evaporation is again *set free as sensible heat.* As an increase of temperature and a diminution of pressure promote evaporation, so a diminution of temperature and an
increase of pressure promote condensation.

194. *Distillation.* — *Distillation* consists in *boiling a liquid
and condensing the vapor* evolved. It enables us to separate
a liquid from the solid matter dissolved in it, and to effect a
partial separation of the more volatile constituents of a mixture
from the less volatile. The apparatus employed is called a *still.*
One of its simpler forms, suitable for distilling water (Figure
134), consists of a retort *a,* the neck of which *c* communicates
with a spiral tube *dd,* called the *worm,* placed in a vessel *e,* containing cold water. The water in the retort is boiled, the steam

9

given off is condensed in the worm, and the *distilled water* is collected in the vessel *g*.   As the condensation proceeds, the water of the cooler becomes heated, and must be renewed.   For this purpose a tube descending to the bottom of the cooler is sup-

Fig. 134.

plied with a continuous stream of cold water from above, while the warm water, which rises to the top, flows out by the tube *i*.

195: *The Spheroidal State.* — This is a peculiar *condition assumed by liquids when exposed to the action of very hot metals.*

If we let fall a drop of water upon a smooth plate of iron or

Fig. 135.

silver, the drop will evaporate more rapidly as the temperature of the plate is increased up to a certain point.   When the temperature exceeds this limit, which, for water, appears to be about 300°, the drop assumes a spheroidal form, rolls about like a ball or spins on its axis, and frequently exhibits a beautiful rippling (Figure 135).   While in this condition it evaporates much more slowly than when the plate was at a lower temperature.   If the plate is allowed to cool, a moment arrives when the globule of water flattens out, and boils rapidly away with a hissing noise. If the temperature of the *liquid* is measured by means of a thermometer with a very small bulb, it is always found to be *below the boiling-point.*

In the spheroidal state *the liquid and the metal plate do not come into contact.* To prove this, the plate used must be quite smooth and accurately levelled. When it is heated, a little water is poured upon it and assumes the spheroidal state. By means of a fine platinum wire the globule is kept at the centre of the plate. It is then very easy, by placing a light behind the globule, to see distinctly the space between the liquid and the plate (Figure 136).

This separation is maintained by *the rush of steam from the under surface of the globule,* which is also the cause of the pe-

Fig. 136.

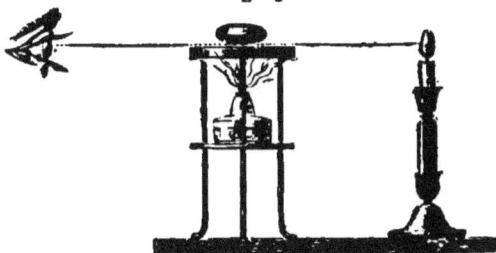

culiar rippling movements. In consequence of the separation, heat can pass to the globule only by radiation, and hence its comparatively low temperature.

D. Measurement of Heat.

196. *The Unit of Heat.* — The temperature of a body indicates its thermal condition, but not the amount of heat in it. The thermometer shows a pound of iron and ten pounds of iron to be of the same temperature, when, of course, the latter has ten times as much heat in it as the former. In the measurement of heat we need some unit in which amounts of heat can be expressed. The English *unit of heat* is the amount of heat required *to raise one pound of water at 32° one degree in temperature.*

197. *Specific Heat.* — If equal bulks of water and of mercury are exposed to the same source of heat, it will be found that the temperature of the mercury will rise faster than that of the water, though the mercury is more than 12 times as heavy as the

water. It has been found that *it requires very different amounts of heat to raise the same weight of different substances one degree in temperature.*

The *specific heat* of a substance is the amount of heat required *to raise one pound of it one degree in temperature.* The specific heat of water is 1, and it is higher than that of any other substance, with the single exception of hydrogen.

198. *A Body in Cooling* 1° *gives out just as much Heat as it takes to Heat it* 1°. — Boil a quarter of a pound of water in a beaker, and the bulb of a thermometer plunged into it will indicate a temperature of 212°. Remove the beaker from the source of heat, and pour the water into another beaker containing a quarter of a pound of water at a temperature of 70°. Stir the mixture a short time with the bulb of a delicate thermometer, and the temperature will be found to be 141°. The first quarter of a pound of water has then lost 71°, and the second has gained 71°; in other words, *the first in cooling one degree has given out just heat enough to warm the second one degree.* The same is true of all other bodies.

199. *The Water Calorimeter.* — A *calorimeter* is an instrument *for measuring quantities of heat.* The *water calorimeter* is a vessel containing water into which a heated substance may be introduced. As the substance cools it imparts some of its heat to the water, and the amount of heat given up by the substance may be calculated from the weight of the water in the calorimeter and the number of degrees the temperature is raised. The number of units of heat received by the water will be equal to the product of the rise of temperature in degrees and the weight of the water in pounds. This method of measuring heat is called the *method of mixture.*

200. *The Latent Heat of Water.* — By the *latent heat of water* we mean the *amount of heat required to melt a pound of ice.* This is found to be 143 units.

201. *The Ice Calorimeter.* — Another method of finding specific heat is *by melting ice.* The substance is first weighed,

then heated to a certain temperature, as 100°, and placed in the vessel M (Figure 137). This vessel is placed within the vessel A, the space between the two being filled with ice. The vessel A is placed in another, B, from which it is also separated by ice. Since the vessel A is surrounded by ice, the heat which melts the ice within it must come wholly from the vessel M. As the ice in A melts, the water runs off through the pipe D. As we know how much heat is required to melt one pound of ice, we need only know *how much ice is melted* by any substance

Fig. 137.

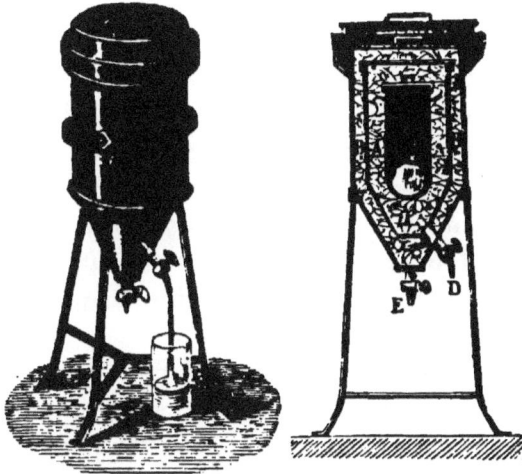

within the box M, in order to find how many units of heat it has given up. *Dividing this by the weight of the substance and by the number of degrees it has cooled,* we get its *specific heat.*

202. *The Latent Heat of Steam.* — The latent heat of watery vapor, or steam, is *higher than that of any other vapor,* being 967 units.

The latent heat of steam may be found by allowing a quantity of steam to pass into a water calorimeter. The steam will be condensed, and the water formed will be cooled to the resulting temperature of the water in the calorimeter. The heat *given out in this condensation and cooling* will raise the temperature of the water in the calorimeter. The amount of this heat may be calculated, as well as the amount of heat *given out in the cooling of the water formed from the steam.* The *difference between these two amounts* will be the amount of heat set free in the condensation of the steam. This, *divided by the weight of the steam,* will give its latent heat.

II. RELATIONS BETWEEN HEAT AND WORK.

203. *Heat consumed in the Performance of Work.* — In expansion, liquefaction of solids, and evaporation, the molecules are always pushed into new positions against some kind of resistance, either internal or external; that is to say, *work is done upon the molecules.* This work is always done *at the expense of heat,* either of that already in the body or of that communicated to the body. Hence, whenever any of these kinds of work are done without the application of heat to the body, some of the heat in the body is consumed and *its temperature falls;* and whenever the work is done by the application of heat, *the temperature of the body rises less than it would* with the same application of heat were no work done.

204. *Heat consumed in Expansion.* — If a thermometer bulb is introduced into the receiver of an air-pump through an opening into which it is fitted air-tight by means of a rubber cork, and the pump is worked, as the exhaustion proceeds the air in the receiver will expand more and more, and the mercury in the stem of the thermometer will fall several degrees, indicating a reduction of temperature. *The air is always chilled when any expansion takes place in it without the application of heat.*

It takes 6.7 units of heat to raise the temperature of a cubic foot of air 490° when the air is confined so that it cannot expand, and 9.5 units to raise the temperature the same amount when the air is free to expand. In the latter case the air will expand enough to double its volume (171). So that 2.8 units of heat are consumed in expanding a cubic foot of air enough to double its volume. The *heat consumed in expansion* is called the *latent heat of expansion.* The conversion of sensible into latent heat is simply the transformation of kinetic into potential energy. When the air *contracts* again, the potential energy is transformed again into kinetic energy, and the *latent heat again becomes sensible.*

205. *Heat consumed in Liquefaction.* — Place some pulverized nitrate of ammonia in a small beaker glass, add an equal bulk of water, and stir the mixture with the bulb of a thermometer. The solid will be rapidly dissolved, and the temperature of the mixture will quickly fall 40 or 50 degrees. If put upon a wet board, the beaker will be quickly frozen to it. In the *liquefaction of a solid* a part of its kinetic energy is transformed into potential energy, and *sensible heat becomes latent heat.* In the *solidification of the liquid* the potential energy is transformed back again into kinetic energy, and the *latent heat again becomes sensible heat.*

In the melting of a solid all the kinetic energy that enters the body is transformed into potential energy by the conversion of the solid into a liquid, and hence there is *no rise of temperature while the solid is melting.*

Fig. 138.

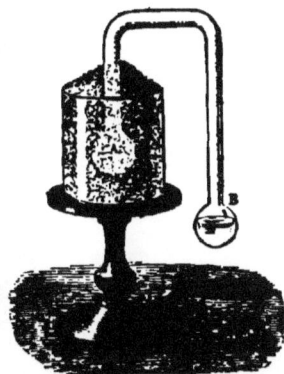

206. *Heat consumed in Evaporation.* — The consumption of heat in evaporation may be illustrated by means of the *cryophorus* (Figure 138). It consists of a bent tube with a bulb at each end. It is partly filled with water and hermetically sealed while the liquid is boiling, thus expelling the air. When an experiment is to be made, all the liquid is passed into *B*, and *A* is plunged into a freezing mixture, or into pounded ice. The cold condenses the vapor in *A*, and thus produces rapid evaporation of the water in *B*. Needles of ice soon appear on the surface of the liquid.

Pour a little water into a small test-tube, and place the tube in a wineglass of ether (Figure 139); then blow a current of air through the ether by means of a pair of bellows. The rapid evaporation of the ether will reduce the temperature sufficiently to freeze the water in the tube in a short time.

In evaporation as in liquefaction, the conversion of sensible into latent heat is merely the transformation of kinetic energy into potential energy.

207. *Freezing Mixtures.* — The ordinary *freezing mixture* is a mixture of *salt and ice.* The salt causes some of the

Fig. 139.

ice to liquefy, and this liquefaction consumes so much heat that the temperature of the mixture is reduced sufficiently to freeze cream within a can which is surrounded by the mixture.

A mixture of solidified carbonic acid and ether, in the receiver of an air-pump from which the air has been exhausted so as to promote the evaporation, evaporates with very great rapidity, and the consumption of heat is so great as to reduce the temperature of the mixture to — 166° F.

A mixture of solidified nitrous oxide and bisulphide of carbon, under similar circumstances, evaporates still more rapidly, and reduces the temperature to — 220° F.

208. *Solidification of Gases.* — If any gas is liquefied by the combined action of cold and pressure, and then allowed to escape into the atmosphere in a fine stream, so as to *evaporate freely*, the temperature will be reduced to such an extent that *a portion of the vapor will be frozen*, so that the gas can be obtained *in a solid state.*

In the case of hydrogen, and some other gases, which cannot be liquefied by the direct action of cold and pressure, if the gas

Fig. 140.

is reduced to the greatest possible degree of density by the combined action of cold and pressure, and then is allowed to expand by a sudden removal of the pressure, *the sudden expansion chills the gas sufficiently to freeze a portion of it* (204). Hydrogen frozen in this way is heard to rattle like hail when it falls on the table.

Faraday was the first to conduct methodical experiments in the liquefaction of gases. The apparatus employed by him (Figure 140) consists of a very strong bent glass tube, closed at both ends. One end of this contains the ingredients which, on the application of heat, evolve the gas to be tried, while the other is immersed in a freezing mixture. The pressure produced by the evolution of the gas in large quantity in a confined space combines with the cold of the freezing mixture to produce lique-faction of the gas, and the liquid collects in the cold end of the tube.

209. *Mechanical Equivalent of Heat.* — Meyer found *the equivalent of a unit of heat in foot-pounds*, by converting heat into mechanical energy through the expansion of air. In the expansion of air the work done is wholly external, namely, that of pushing aside the surrounding air. We have seen (204) that it takes 2.8 units of heat to expand a cubic foot of air to double its volume. To ascertain the amount of work done in pushing away the surrounding air, Meyer imagined his cubic foot of air at the bottom of a prismatic box whose section was a foot square, so that the air could expand only upward. The upper surface of the cubic foot of air contains 144 square inches. Hence the weight of the column of air pressing upon this surface is about $144 \times 15 = 2160$ pounds ; and when the cubic foot of air expands so as to double its volume, this weight must be raised one foot high. Hence 2.8 units of heat are equivalent to 2160 foot-pounds of mechanical energy, and *one unit of heat is equivalent to 772 foot-pounds.* This is known as the *mechanical equivalent of heat.*

210. *The Steam-Engine.* — The molecular energy of heat can be made to do mechanical work by means of the arrangement shown in Figure 141. The steam derives its expansive power from the heat, and this expansive power is made to work a piston in the cylinder of the steam-engine. The steam from the boiler passes through the tube *x* into the *steam-box d*. Two pipes run from this box, one *a* to the top and the other *b* to the bottom of the cylinder. A sliding-valve *y* is so arranged as always to close one of the pipes to the steam-box and open it to the *exit-pipe O*, and, at the same time, to open the other pipe to the steam-box and close it to the exit-pipe. In the right-hand figure the lower pipe *b* is open, and the steam can pass in under the piston and force it up. At the same time the steam which has done its work on the other side of the piston passes out from the cylinder through the pipes *a* and *O*.

The sliding-valve is connected by means of the rod *i* with the crank of the engine, so that it moves up and down as the piston moves down and up. As soon, then, as the piston has reached the top of the cylinder, the sliding-valve is brought into the position shown in the left-hand figure. The steam now passes into the cylinder above the piston through the pipe *a*, and forces the

Fig. 141.

piston down, and the steam on the other side which has done its work goes out through *b* and *O*. The sliding-valve is now again in the position shown in the right-hand figure, and the piston is driven up again as before ; and thus it keeps on moving up and down, or in and out.

III. DISTRIBUTION OF HEAT.

A. Conduction.

211. *Illustration of Conduction.* — If heat is applied to one end of a bar of metal, it is slowly propagated through the substance of the bar, producing a rise of temperature

which is first perceptible near the heated end, and afterwards in more remote portions. The *transmission of heat
from molecule to molecule* through the substance of the body
is called *conduction.*

If the application of heat to one end of the bar is continued
for a sufficiently long time, and with great steadiness, the different portions of the bar will at length cease to rise in temperature, and will retain steadily the temperatures which they have
acquired. We may thus distinguish two stages in the experiment : 1st, the *variable* stage, during which all portions of the
bar are rising in temperature ; and 2d, the *permanent* state,
which may subsist for any length of time without alteration. In
the former, the bar is gaining heat ; that is, it is receiving more
heat from the source than it gives out to surrounding bodies.
In the latter, the receipts and expenditure of heat are equal, not
only for the bar as a whole, but for every small portion of it.

In the permanent state no further accumulation of heat takes
place. All the heat which reaches an internal particle is transmitted by conduction, and the heat which reaches a superficial
particle is given off partly by radiation and air-contact, and
partly by conduction to colder neighboring particles. In the
earlier stage, on the contrary, only a portion of the heat received
by a particle is thus disposed of, the remainder being accumu-
lated in the particle, and serving to raise its temperature.

212. *Conducting Power of Solids.* — Different solids are
found to vary much in *conductivity,* or conducting power.

The following experiments are often adduced in illustration
of the different conducting powers of different solids.

Fig. 142.

Two bars of the same size, but of different materials (Figure
142), are placed end to end, and small wooden balls are attached

by wax to the under surfaces at equal distances. The bars are then heated at their contiguous ends, and, as the heat extends along them, the wax melts, and the balls successively drop off. If the heating is continued till the permanent state arrives, it may generally be concluded that the bar which has lost most balls is the best conductor.

The apparatus shown in Figure 143 consists of a copper box having on one side a row of holes in which rods of different materials can be fixed. The rods having been pre-viously coated with wax, the box is filled with boil-ing water, which comes into contact with the inner ends of the rods. The

Fig. 143.

wax gradually melts as the heat travels along the rods; and if the experiment is continued till the melting reaches its limit, those rods on which it has extended furthest are, generally speaking, the best conductors. It is thus found that different *metals* are *not equally good conductors* of heat, and that the more familiar ones may be arranged in the following order, beginning with the best conductors : *Silver, copper, gold, brass, tin, iron, lead, platinum, bismuth.*

Metals, though differing considerably one from another, are as a class greatly *superior in conductivity to other substances*, such as wood, marble, brick, etc. This explains several familiar phenomena. If the hand is placed upon a metal plate at the temperature of 50°, or plunged into mercury at this temperature, a very marked sensation of cold is experienced. This sensation is less intense with a block of marble at the same temperature, and still less with a piece of wood. The reason is that the hand, which is at a higher temperature than the substance to which it is applied, gives up a portion of its heat, which is *con-ducted away by the substance;* consequently a larger portion of heat is parted with in the case of the body of *greater conducting power.*

213. *Conducting Power of Liquids.* — With the exception of mercury and other melted metals, *liquids are exceedingly*

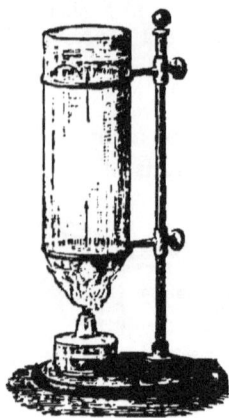

bad conductors of heat. This can be shown by heating the upper part of a column of liquid, and observing the variations of tempera-ture below. These will be found to be scarcely perceptible, and to be very slowly produced. If the heat were applied below (Figure 144), we should have the process called *convection* of heat; the lower layers, made lighter by expansion, would rise to the sur-face, and be replaced by colder ones from above, which would be heated and rise in their turn, the *circulation* thus producing a general heating of the liquid. When heat is applied above, the expanded layers remain in their place, and the rest of the liquid can be heated only by conduction and radiation.

Fig. 144.

The following experiment is an illustration of the very feeble conducting power of water. A piece of ice is placed at the bottom of a glass tube (Figure 145), which is then partly filled with water ; heat is applied to the middle of the tube, and the upper portion of the water may be made to boil without melting the ice below.

Fig. 145.

214. Conducting Power of Gases. — Of the conducting power of gases it is almost impossible to obtain any direct proofs, since it is ex-ceedingly difficult to prevent the interference of convec-tion and direct radiation. We know, however, that they are *exceedingly bad conductors.* In fact, in all cases where gases are enclosed in small cavities where their movement is difficult, the system thus formed is a very bad conductor of heat. This is the cause of the feeble conducting powers of many kinds of cloth, of fur, eider-down, felt, straw, saw-dust, etc. Materials of this kind, when used as articles of clothing, are commonly said to be *warm*, because they *hinder the heat of the body from escaping.* If a garment of eider-down or fur were compressed so as to expel the greater part of the air, and to reduce the substance to a thin sheet, it would be found to be a much less warm covering than before, having become a better conductor. We thus see that *it is the presence of air which gives these substances their feeble conducting power,* and we are accordingly justified in assuming that air is a very bad conductor of heat.

B. Convection.

215. *Convection Currents.* — Although liquids and gases are very poor conductors of heat, they allow heat to be *distributed through them readily* by *convection currents.* When heat is applied to any portion of a fluid, the heated portion expands, becomes lighter, and rises, allowing colder portions to take its place and become heated in turn ; that is, the system of currents shown by the arrows in Figure 144 is formed. There will be an upward current at the centre of the heated region, an outflow in every direction above, downward currents on every side, and an inflow from every direction below. It is chiefly in this manner that heat is distributed through liquids and gases.

C. Radiation and Absorption.

216. *Illustration of Radiation.* — When two bodies at different temperatures are brought opposite to each other, an unequal exchange of heat takes place through the intervening distance ; the temperature of the hotter body falls, while that of the colder rises, and after some time the temperature of both becomes the same. This *propagation of heat across an intervening space* is what is meant by *radiation*, and the heat thus transmitted is called *radiant heat.* Instances of heat communicated by radiation are the heat of a fire received by a person sitting in front of it, and the heat which the earth receives from the sun.

217. *Radiations will traverse a Vacuum.* — This last instance shows us that radiation as a means of propagating heat is *independent of any ponderable medium.* But since the solar heat is accompanied by light, it might still be questioned whether non-luminous heat could in the same way be propagated through a vacuum.

This was tested by Rumford in the following way. He con-

Fig. 146.

structed a barometer, the upper part of which was expanded into a globe (Figure 146). A thermometer was hermetically sealed into the top, so that the bulb of the thermometer was at the centre of the globe. The globe was thus a Torricellian vacuum-chamber. By melting the tube with a blow-pipe, the globe was separated, and was then immersed in a vessel containing hot water, when the thermometer immediately rose to a temperature higher than could be due to the conduction of heat through the stem. The heat had therefore been communicated by *direct radiation through the vacuum* between the sides of the globe and the bulb *a* of the thermometer.

218. *Radiant Heat travels in Straight Lines.* — In a uniform medium the radiation of heat takes place *in straight lines.* If, for instance, between a thermometer and a source of heat there are placed a number of screens, each pierced with a hole, and if the screens are so arranged that a straight line can be drawn through the holes from the source to the thermometer, the temperature

of the latter immediately rises ; if a different arrangement is adopted, the heat is stopped by the screens, and the thermometer indicates no effect.

The heat which travels *along any one straight line* is called a *ray* of heat. Thus, we say that rays of heat issue from all points of the surface of a heated body, or that such a body emits rays of heat.

219. *Molecular Theory of Radiation.* — According to the molecular theory, *radiations originate in the vibrations of the atoms within the molecule.* Each kind of atoms seems to have certain characteristic rates of vibration, and when the molecules in their motions come into collision, their atoms are thrown into vibration ; these vibrations are communicated to the surrounding ether (3), and are *propagated through the ether in minute waves and with enormous velocity.* As the temperature of the body rises the agitation of its molecules becomes more energetic, and the more violent collisions of the molecules produce more powerful vibration of the atoms. Hence the radiation becomes more intense as the temperature rises.

220. *Different Kinds of Radiation.* — At low temperatures bodies emit only *obscure* radiations. When the temperature reaches a certain point, the body becomes red-hot, and begins to emit *luminous* radiations. At a still higher temperature it becomes white-hot.

221. *Diathermanous Bodies.* — A body, like air, which will *allow thermal rays to pass readily through* it is said to be *diathermanous.* If a polished plate of glass is held in front of a body heated to dull redness, it will stop nearly all the heat emitted by it. If the same plate of glass is held in front of a body at bright white heat, it will allow considerable heat to pass through it. Glass is diathermanous to luminous radiations, but only slightly so to obscure thermal radiations. A solution of alum is still less diathermanous to obscure thermal rays, although it allows the luminous rays to pass readily through it. A solution of iodine in bisulphide of carbon, on the contrary, is perfectly diathermanous to the obscure thermal rays and per-

fectly opaque to the luminous rays. A polished plate of rock salt is diathermanous to both the obscure and luminous rays. .

222. *The Effect of Rise of Temperature on Radiation.* — If the temperature of a body is gradually raised to the highest possible point, and a cell of the iodine solution is used to cut off the luminous radiations, the obscure thermal radiations will be found to grow more and more intense, both before and after the body begins to emit luminous radiations. A rise of temperature, then, has two effects upon the radiation of a body ; it *causes its obscure radiations to become more intense*, and *gives rise to new radiations*. The latter radiations differ from the former in having quicker vibrations and shorter waves. The *radiations of longest and shortest wave-lengths are obscure*, while *those of medium wave-lengths are luminous*. The *radiations of all wave-lengths are thermal*, but the thermal power is greatest in radiations of long wave-lengths, and least in those of short wave-lengths. *All radiations are capable of producing certain chemical effects*, but the chemical or *actinic* power is least in radiations of long waves and greatest in the short waves. The radiations of bodies have, accordingly, been divided into three classes ; namely, *obscure thermal, luminous*, and *obscure actinic*. At low temperatures bodies emit only the first class of radiations ; at higher temperatures, the first and second classes ; and at still higher temperatures, all three classes.

223. *Absorption.* — *Absorption* is the *reverse of radiation*. When the minute waves of the ether encounter the molecules of gross matter, they throw the atoms into vibration, provided they can vibrate at the same rate as the particles of the ether in the waves. In this way the rays are taken up and absorbed by bodies. It is only those rays which are *absorbed* by a body that *heat* it. Bodies are *not warmed* at all by the rays which they *transmit*.

224. *Good Radiators are Good Absorbers.* — Rough blackened surfaces are *better radiators* than smooth polished surfaces, and they are also *better absorbers*.

This may be shown by the following experiment. Two metallic plates *A* and *B* (Figure 147) of the same size are

mounted on standards which move to and fro on a sliding
bar at the bottom. Between these plates there is a rod for
supporting a ball at the height
of the centre of the plates. *A*
is coated with polished nickel
on both sides, and *B* with nickel
on one side and lampblack on
the other. *B* is made to turn
on its standard so that the surface coated with lampblack may
be turned either towards the
ball or from it. First, turn the
nickel faces of the plate towards the ball, heat the ball to dull
redness, place it upon its rod, and move both plates up against
it so that they may be heated equally. Place a differential thermometer (179) as shown in the figure, so that its bulbs shall be
equally distant from the two plates. One of the bulbs will
be heated by radiation from the nickel surface, and the other
by radiation from the blackened surface. The liquid in the
stem will move towards the former bulb, showing that the latter
bulb is hotter, and that *the radiation is more powerful from the
blackened surface.* Now reverse plate *B*, turning its blackened
face towards the ball, remove the ball, and allow both plates to
cool. Place each plate against one of the bulbs of the thermometer, and arrange them so that they shall be equally distant
from the ball. Heat the ball and replace it on the rod. The
plates will now become heated by absorption of radiations from
the ball. They will receive equal radiations, but the thermometer will indicate that the plate with the lampblack coating
towards the ball is the hotter. Hence *the blackened surface is
the better absorber.*

Different *gases*, as well as different solids and liquids, differ
in their absorptive power and in the kind of rays which they
absorb. *Watery vapor* among gases corresponds to glass
among solids and a solution of alum among liquids. It is diathermanous to luminous rays, but much less so to obscure
rays.

Stoves and radiators, which are designed to give out heat,
should have rough blackened surfaces; while a teapot, which

Fig. 147.

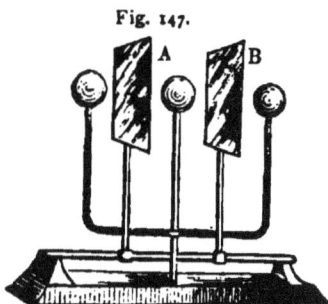

is designed to keep the liquid in it hot, should have a bright polished surface.

225. *Hot-Houses.* — A hot-house is a structure covered with glass. On a sunny day the temperature will be several degrees higher within such a structure than on the outside. The luminous heat which comes from the sun passes readily through the glass and falls upon the objects within. These absorb the heat and in turn send back obscure heat. This heat is stopped by the glass. Hence the heat accumulates within the hot-house. A hot-house may be described as *a trap to catch sunbeams.* Even at night and on a cold cloudy day it will be warmer within a hot-house than on the outside, the glass preventing the obscure radiations from passing off into space. The watery vapor in the atmosphere acts just like the glass of the hot-house.

IV.

LIGHT.

A. RADIATION.

226. *Luminous Bodies*. — Bodies, like a gas-jet or the sun, which *emit light of their own*, are said to be *luminous*. Light is now believed to originate in *extremely minute and rapid vibrations of the atoms* of matter. These vary in rapidity from about 400 million million to about 760 million million a second. The atoms of all luminous bodies are supposed to be vibrating at this enormous rate.

When a body is heated its atoms are thrown into more and more rapid vibration, and when the rate of vibration reaches 400 million million a second the body begins to become luminous. In the case of a candle-flame or gas-jet, these rapid vibrations are produced by the clashing of the atoms of oxygen, hydrogen, and carbon as they rush into combination. A blacksmith may heat a nail red-hot by vigorously hammering it. Each blow of the hammer throws the atoms of the nail into more rapid vibration, till they finally vibrate fast enough to develop light.

227. *Propagation of Light by the Ether*. — As the atoms of matter vibrate in the ether (3) in which they are immersed, they communicate their vibration to it. The vibrations thus started *are propagated through the ether in every direction in minute waves and with an inconceivable velocity*. These ethereal waves vary in length according to the rate of the atomic vibrations. It takes somewhat more than 35,000 of the longest of these waves, and somewhat

less than 70,000 of the shortest, to make the length of an
inch.   The vibrations are transverse, so that each luminous
wave is made up of crest and hollow, like a water wave.
Light and luminous radiations are the same thing.

228. *Velocity of Light.* — The velocity of light is about
186,000 *miles a second.*   It was first determined by Roemer,
a Danish astronomer, by a study of the eclipses of one of
Jupiter's moons.   He found, by examining a long series of
observations, that the mean interval between two successive
eclipses of the moon was about 42½ hours, but that the
interval varied according to the motion of the earth with
respect to Jupiter.   When the earth was moving away

Fig. 148.

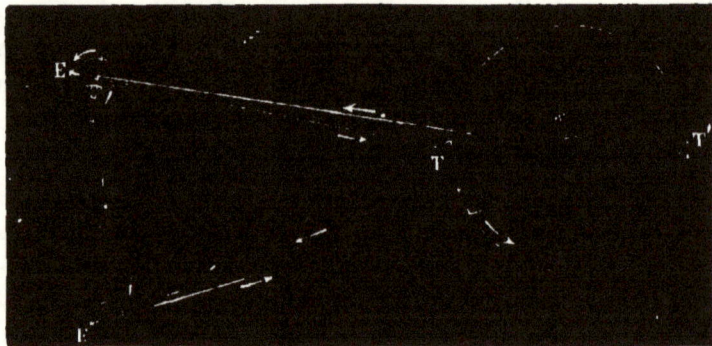

from Jupiter from T to T' (Figure 148), the intervals
were longer than the mean, till at T' the eclipse occurred
about 16½ minutes late ; when the earth was moving
towards Jupiter, from T' to T, the intervals were shorter
than the mean.   Now we cannot be aware of the eclipse
till the light which left the moon just as it entered Jupiter's
shadow has reached the earth ; and the distance this light
has to travel is continually increasing as the earth travels
from T to T', and decreasing as the earth travels from T'
to T.   Roemer concluded that this must be the reason
why the intervals between the eclipses were longer than
the mean in the one case and shorter in the other.   As the

eclipse occurred 16½ minutes late at T', he concluded that it must take light about 16½ minutes to cross the earth's orbit. As this distance is about 184,000,000 miles, light must travel at the rate of about 186,000 miles a second. This velocity would carry light around the earth in about ⅛ of a second.

Great as is this velocity, it is believed that the nearest fixed star is so distant that it would take light over three years to reach us from it, while the most distant stars are, at least, a thousand times more remote. Were all the stars in the heavens blotted out of existence to-night, it would be over three years before we should miss any of them, a quarter of a century before we should miss many, and thousands of years before we should lose them all. The light which will enter our eyes as we glance at some star to-night probably started on its journey before the building of the great pyramids, and has been travelling eight times the distance around the earth every second since.

Fig. 149.

229. *Rectilinear Propagation of Light.* — When sunlight enters through an opening into a darkened room, it illumines the dust in the atmosphere in its path, which may then be easily traced. This path is always found to be

straight.   Light always *traverses a homogeneous medium in straight lines*.   A *single line of light* is called a *ray*, and a *collection of rays* a *beam*.

230. *Images produced by Small Apertures.* — If a white screen is placed opposite a small opening in a shutter of a darkened room, *an inverted picture of the outside landscape* will be formed on the screen (Figure 149). The smaller the opening, the sharper the image.

The formation of this image is due to the rectilinear propagation of light. The point A (Figure 150) is sending out rays in all directions in straight lines. The rays from this point which pass through the small opening must fall upon A' of the screen. In the same way the rays from B must fall upon B'. As A sends light to no part of the screen except A', and as A' receives light from no part of the object but A, the color and brightness of the spot A' will depend upon the color and brightness of A; in other words, A' will be the *image* of A. In like manner B' will be the image of B, while the points of the object between A and B will have their images at corresponding points between A' and

Fig 150.

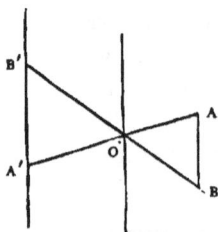

Fig. 151.

B'. An inverted image of $A B$ will thus be formed between A' and B'.

Hold a card with a large pin-hole in it between a candle and a screen (Figure 151), and an inverted image of the candle will be formed on the screen.

Fig. 152.

When the sun shines through a small hole into a room with the blinds closed, whatever may be the shape of the opening, the image of the sun formed on the floor or wall will be round or oval, according as it falls upon a surface which is perpendicular or oblique to the rays (Figure 152). When the sun shines through the foliage of trees, the spots of light on the ground will always be round or oval, whatever may be the shape of the openings through which the light comes, provided they are sufficiently small. When the sun is undergoing eclipse, the progress of the eclipse may be watched by noticing the shape of these spots, which will always be that of the uneclipsed portion of the sun's disc.

231. *Shadows*. — Bodies which, like glass, *allow light to pass readily through* them, are said to be *transparent*. Bodies which *do not allow light to pass through* them are said to be *opaque*.

Owing to the rectilinear propagation of light, opaque bodies in front of a light must necessarily *cast shadows*, that is, *shut off the light from some of the space behind them*.

If the luminous body S (Figure 153) is a *mere point*, the body M will cast a *well-defined* shadow GH upon the screen PQ. If the straight line SG is kept fast at S, and carried round the sphere M, touching it all the time, it will describe a *cone*.

Fig. 153.

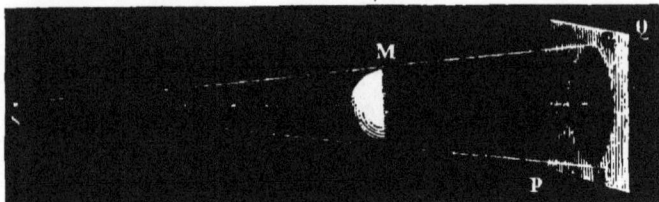

The part MG, as it passes round, will exactly mark the limits of the shadow cast by M.

If the luminous body is *not a mere point*, the shadow of M (Figure 154) upon the screen will be *indistinct in outline*.

Prolong the line GS to A. Keep the point A fixed and carry the line AG around the spheres S and M, keeping it in contact with both. The line will describe a cone, and the part

Fig. 154.

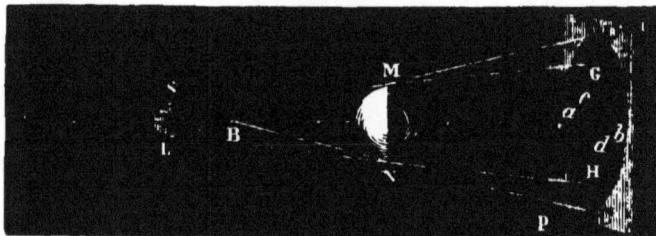

MG will mark out the space from which *the light is entirely excluded*. This is called the *umbra* of the shadow. If the line SC is kept fixed at B, and then carried round the two spheres, it will describe a *double* cone, whose apex will be at B. The part NC of this line will mark the extreme limits of the shadow. From the part outside of the umbra only *a portion of the light is excluded*, and the farther we pass from the umbra the less the light excluded. This part of the shadow is called the *penumbra*. It will be seen at once from the figure, that the light from S will reach all the space between D and G, and the light from L all the space between C and H.

232. *Illumination.* — The *illuminating power* of a source of light *diminishes as the square of the distance from the illuminating body increases.*

In Figure 155 the disc CD is held parallel with the screen AB, and half-way between the screen and the source of light

Fig. 155

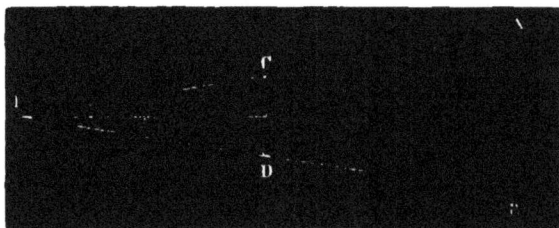

L. The diameter of the shadow on the screen will be twice that of the disc, and the area of the shadow four times that of the disc. The disc receives all the light that would fall upon the space covered by the shadow, were the disc removed. Hence the illumination of the disc is four times as intense as that of the screen. If the disc were held one third of the way from L to the screen, the area of the disc would be one ninth that of its shadow, and the illumination of the disc would be nine times as intense as that of the screen.

Fig. 156.

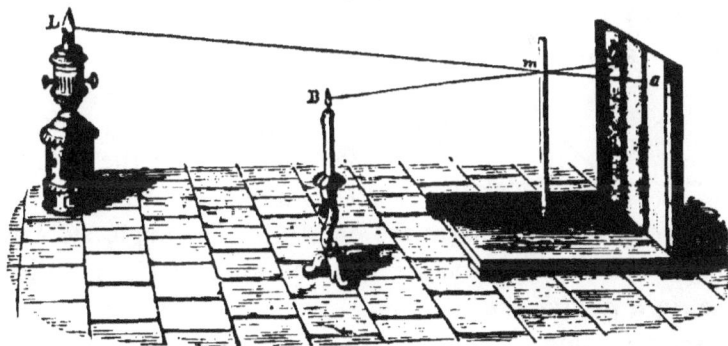

233. *Photometry.* — *Photometry* is the *measurement of the relative illuminating power of different sources of light ;* and the instrument used is called a *photometer.*

Rumford's photometer, based upon the *comparison of shadows*, is one of the simplest of these instruments. An opaque rod *M* (Figure 156) is placed in front of a ground-glass screen. The lights *L* and *B* to be compared are placed so that each casts a separate shadow of the rod upon the screen. These distances are then made such that the two shadows *a* and *b* are of exactly the same intensity. The screen must then be receiving the same illumination from each light; for the shadow cast by *B* is illumined by *L*, and that cast by *L* is illumined by *B*. Hence the illuminating power of the two lights will be to each other *as the squares of the distances of the lights from the screen.*

B. REFLECTION.

234. *Diffusion.* — When light meets the surface of a new medium, a portion of it is *diffused*, that is, *thrown back and scattered irregularly in every direction.* It is by means of the light thus diffused that we are enabled to see the surfaces of non-luminous bodies. Smooth polished surfaces diffuse less light than rough irregular ones, but the most highly polished mirror diffuses enough light to enable us to see its surface, though sometimes with difficulty.

235. *Reflection.* — On meeting the surface of a new medium, a portion of the light is *reflected*, that is, *thrown back in a definite direction.* In Figure 157

Fig. 157.

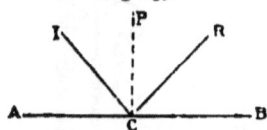

A B represents the surface of the new medium, *I C* the ray coming to it, or the *incident* ray, and *C R* the *reflected* ray. *P C* is a perpendicular to the surface of the medium at the point *C*. The angle *I C P* is called the *angle of incidence*, and the angle *R C P* is called the *angle of reflection.*

In reflection, the *angles of incidence and reflection are always equal.* The smoother the surface of a medium, the greater the proportion of the light reflected from it. *Good reflecting surfaces* are called *mirrors*.

236. *Images formed by Plane Mirrors.* — It is *by reflected light* that we see *images* of objects in reflecting surfaces. These *visible images* formed by reflection correspond to the *echoes* formed by reflection in the case of sound.

Figure 158 represents a pencil of rays emitted from the highest point of a candle-flame to the eye of an observer. The rays have exactly the same degree of divergence after reflection as before, and if prolonged backward would meet just as far behind the mirror as the point from which they come is in front of it. The same would be true of the rays coming from every point of the object. Hence *an image seen in a plane mirror will seem just as far behind the mirror as the object is in front of it.* This is not only true of the image as a whole, but also of each part of it.

Fig. 158.

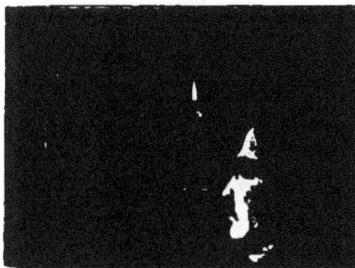

Fig. 159.

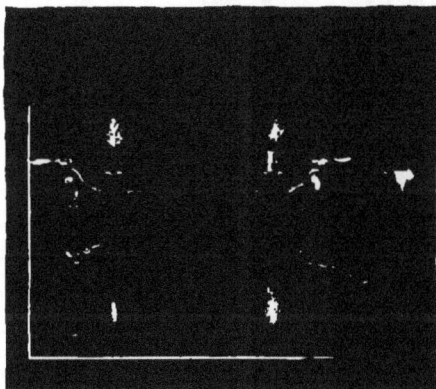

237. *Images formed by two Mirrors at an Angle to each other.* — Figure 159 shows the images that would be formed by two mirrors at right angles to each other, one being horizontal and the other vertical.

Figure 160 shows the images that would be formed if an object were placed between two mirrors facing each other at an

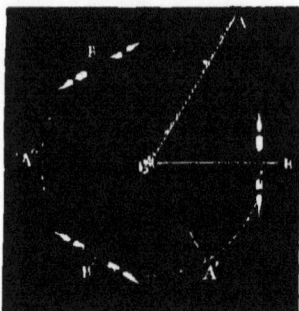

Fig. 160.

angle of 60°. When the mirrors are inclined to each other, the images formed are always arranged in the circumference of a circle, whose centre is at the intersection of the mirrors, while its circumference passes through the object.

238. *The Kaleidoscope.* — The *kaleidoscope* is an optical toy, invented by Sir David Brewster. It consists of a tube containing two glass plates, extending along its whole length, and inclined at an angle of 60°. One end of the tube is closed by a metal plate, with a hole in the centre, through which the observer

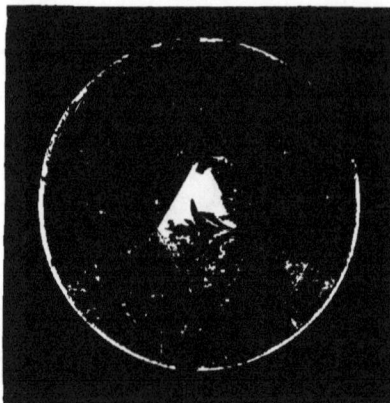

Fig. 161.

looks in; at the other end there are two plates, one of ground and the other of clear glass (the latter being next the eye), with a number of little pieces of colored glass lying loosely between them. These colored objects, together with their images in the mirrors, form *symmetrical patterns of great beauty*, which can be varied by turning or shaking the tube, so as to cause

the pieces of glass to change their positions (Figure 161).

C. REFRACTION.

239. *Refraction.* — If a beam of light is allowed to fall obliquely upon water, it will be seen to be bent on entering the water, though it will continue to move on in a straight line after it has passed into the water. This *bending of a ray of light, in passing obliquely from one medium to another*, is called *refraction*.

If a coin or other object *m n* (Figure 162) is placed on the bottom of a vessel with opaque sides, so as just to be concealed from an eye at *O*, and the vessel is then filled with water, the bottom of the vessel will seem to rise and the object will come into view. This is because

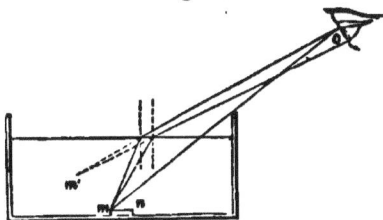

Fig. 162.

the pencils of rays coming from the object at *m* will be suddenly bent on entering the air and will reach the eye as if they came from *m'*, where the object will appear to be.

For a similar reason, a stick partly immersed in water, in an oblique position, will appear bent, as shown in Figure 163.

Fig. 163.

When a ray of light passes obliquely *from a rarer into a denser medium*, it is bent *towards a perpendicular drawn to the surface of the medium* at the point of contact of the ray.

In Figure 164 *A B* represents the surface of a denser medium, *I C* the incident ray, *C R* the refracted ray, and *P C H* a perpendicular to the surface of the medium at the point *C*. The

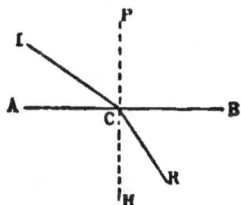

Fig. 164.

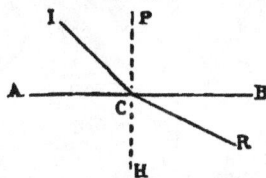

Fig. 165.

angle *R C H* is the *angle of refraction.* In this case the angle of refraction is *less* than the angle of incidence.

When a ray of light *enters a rarer medium* obliquely, it is bent *from a perpendicular to the surface of the medium* at the point of contact.

In Figure 165 $A B$ represents the surface of a rarer medium, $I C$ the incident ray, $C R$ the refracted ray, and $P C H$ the perpendicular. In this case the angle of refraction is *greater* than the angle of incidence.

When a ray of light enters any medium *perpendicularly*, there is *no refraction*.

240. *Total Reflection.* — The angle of incidence may

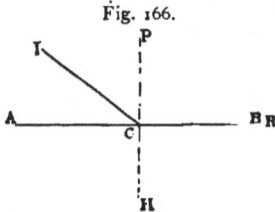

Fig. 166.

have any value from 0° up to 90°. When light enters a *denser* medium, the angle of refraction is *less* than the angle of incidence, and hence *always less than* 90°. But when light enters a *rarer* medium, there is always a certain angle of incidence $I C P$ (Figure 166) at which the angle of refraction $H C R$ is *equal to* 90°.

Fig. 167.

This angle is called the *limiting angle,* or the *critical angle.* When the media are air and water, this angle is about 48½ degrees. For air and the different kinds of glass it ranges from 38° to 41°.

When the angle of incidence *exceeds the limiting angle,* none of the light will enter the medium, however transparent it may be. In this case the light will be *totally reflected,* the angle of reflection being equal to that of incidence.

If a glass of water, with a spoon in it, is held above the level of the eye (Figure 167), the under side of the surface of the water is seen to shine like polished silver, and the lower part of the spoon is seen reflected in it. The rays of light which pass upward through the water at a certain angle are *totally reflected* on meeting the air.

D. DISPERSION.

241. *The Dispersion Spectrum.* — If a glass prism (Figure 168) is held with its edge down to the path of a thin beam of light, the spot of light on the screen will be raised

,Fig. 168.

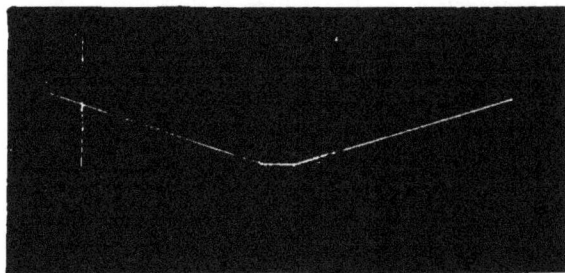

and be changed into a beautifully colored band, in which the colors are arranged in the order of *red, orange, yellow, green, blue, indigo, and violet.* The *colored band produced by the passage of a beam of light through a prism* is called the *dispersion spectrum.* The raising of the spot of light on the screen is due to the bending of the beam as a whole by the prism ; and the formation of the colored band, to the unequal bending of the different colored rays of which white light is composed, red being bent the least and violet the most of all the rays. The *separation of the colored rays by refraction* is called *dispersion.*

The *refrangibility* of light is found to depend upon *the length of its waves ;* the shorter the waves, the more refrangible the ray. The violet rays are more refrangible than the red because they have shorter waves.

11

In the case of sunlight and of light from any intense source of heat, it is found that the *thermal* or heating power of the spectrum extends considerably beyond the red, and the *actinic* or chemical power considerably beyond the violet. The complete spectrum is composed of three parts, a *luminous* portion at the centre, an *obscure thermal* portion beyond the red, and an *obscure actinic* portion beyond the violet. Every portion of the spectrum is *thermal*, but the thermal power increases rapidly as we approach the red end, and is greatest just beyond the red. Every part of the spectrum is also *actinic*, but the greatest actinic power is in the region of the blue. Only the central part of the spectrum is *luminous*, and the greatest luminosity is in the region of the yellow and green.

242. *Achromatic and Direct-Vision Prisms.* — The refractive power of a substance is independent of its dispersive power. Hence, by using different kinds of glass, it has been found possible to construct prisms which shall have equal refractive powers and unequal dispersive powers, or equal dispersive and unequal refractive powers. If two prisms of crown and flint glass are constructed so as to have equal powers of bending a beam of light as a whole, the flint-glass prism will produce greater dispersion than the crown-glass. If, on the other hand, the two prisms are constructed so as to produce equal dispersion, the crown-glass prism will bend the ray as a whole more than the flint-glass.

When two prisms of *equal dispersive* and *unequal refractive* powers are combined, with the thicker part of one beside the thinner part of the other (Figure 169), they form what is called

Fig. 169.

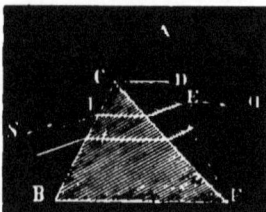

an *achromatic prism*. Such a prism produces *refraction without dispersion*. *Achromatic* means *without color*.

When two prisms of *equal refractive and unequal dispersive* powers are combined as above, they form what is known as a *direct-vision prism*. Such a prism produces *dispersion without refraction*. In using it we look directly at the object, while with any other prism we are obliged to look somewhat away from the object (Figure 170).

Fig. 170.

243. The Spectroscope. — The *spectroscope* is an instrument for *examining spectra*. A *simple* spectroscope is shown in

Fig. 171.

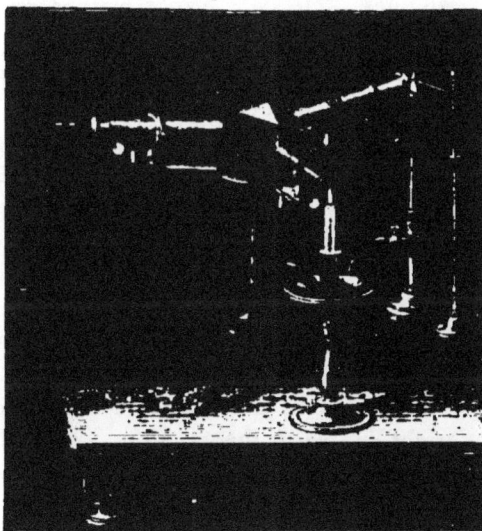

Figure 171. The tube at the right is called the *collimator* tube.

The light to be examined is admitted through a narrow opening at the end of the tube, and the rays are rendered parallel by means of a lens within it. The light is then dispersed by the prism, and the spectrum examined by means of the telescope at the left of the prism. The tube in front of the prism has a scale engraved on glass in the opening at the end next to the candle. The light from the candle which shines through this scale is reflected from the side of the prism into the telescope, so as to form an image of the scale alongside that of the spectrum.

244. *Three Kinds of Spectra.* — On examining with the spectroscope the light from an incandescent solid, its spectrum will be found to be a *continuous band of colors*, changing by insensible gradations from red at one end to violet at the other. Such a spectrum is called a *continuous* spectrum. *Incandescent solids and liquids give continuous spectra.*

If we examine with the spectroscope the light from luminous strontium vapor, its spectrum (see frontispiece) will be seen to be made up of *bright lines and dark spaces.* Such a spectrum is called a *bright-lined* or *broken* spectrum. *Vapors and gases, when luminous, give bright-lined spectra.* The spectra of different gases and vapors differ in the number and position of these lines. Hence a vapor may be recognized by its spectrum.

The dark spaces of these spectra are due to the absence of certain rays. While incandescent solids and liquids emit rays of all degrees of refrangibility, luminous vapors and gases emit those only of particular degrees of refrangibility. Each vapor or gas emits just as many sets of rays as there are bright lines in its spectrum. The number of these lines ranges from one up to several hundred.

The *analysis of light by means of the spectroscope* is called *spectrum analysis.* The spectrum of an incandescent solid or liquid, when shining through a luminous vapor or gas, is made up of *dark lines* separated by bright spaces, there being a dark line for every bright line which the gas alone would give. Such spectra are called *reversed spectra, the spectrum of the gas being reversed by the light of the solid which passes through it.*

E. LENSES.

245. *Forms of Lenses.* — A *lens* is *a transparent medium having at least one curved side.* Lenses are usually made of glass, and are circular in outline. Their curved surfaces are usually spherical. They are divided into two classes, according to their shape, namely, *convex lenses* and *concave lenses.* Every convex lens has at least one *convex* surface, and is *thickest at the centre;* and every concave lens has at least one *concave* surface, and is *thickest at the margin.* There are three forms of each class of lenses. These six forms of lenses are shown in section in Figure

Fig. 172.

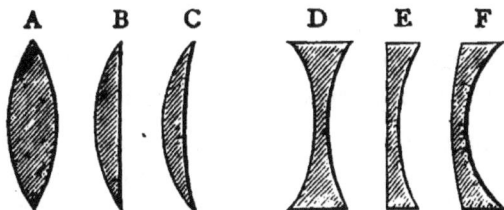

172. The first three are convex and the last three concave lenses. *A* is a *double-convex* lens, having two convex surfaces. *B* is a *plano-convex* lens, having one plane and one convex surface. *C* is a *concavo-convex* lens, having a concave and a convex surface, the convex surface having the greater curvature. This lens is often called a *meniscus.* *D* is a *double-concave* lens, having two concave surfaces. *E* is a *plano-concave* lens, having a plane and a concave surface. *F* is a *convexo-concave* lens, having a convex and a concave surface, the concave surface having the greater curvature.

246. *The Optical Centre of a Lens.* — There is for every lens *a point, any straight line drawn through which will meet on opposite sides of the lens portions of surface which are parallel.* This point is called the *optical centre* of the lens.

247. *Axes and Foci of Lenses.* — *Any straight line drawn through the optical centre* of a lens is called an *axis.* An axis which passes *through the centre of curvature* of a lens is called the *principal* axis, and every other axis a *secondary* axis.

Every ray of light which coincides with an axis will emerge from a lens with the same direction it had before entering, since it will pass through a portion of a medium having parallel sides. Every other ray which passes through a lens will be deflected towards the thicker part of the lens. In the case of a *convex* lens the deflection will be towards the *centre* of the lens, and of a *concave* lens towards the *margin.*

When the rays, on emerging from a lens, are either convergent or divergent, *the points towards which they converge or from which they diverge* are called *foci.* When the rays are *convergent* on emerging from the lens, the focus is *real;* and when they are *divergent,* it is *virtual.*

248. *Parallel Rays with Lenses.* — *Parallel* rays with a convex lens (Figure 173) *become convergent* on emerging from the lens, and have a *real focus on the opposite side of the lens* to that on which they enter and *on the axis to which the rays are parallel.*

Fig. 173.

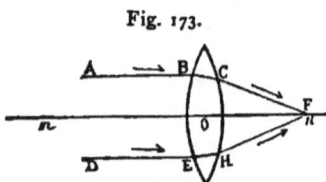

Parallel rays with a *concave* lens (Figure 174) *become di-*

Fig. 174.

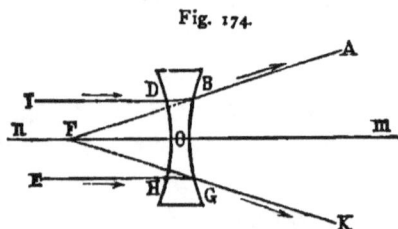

vergent, and have a *virtual focus on the same side of the lens* as that on which the rays enter and *on the axis to which the rays are parallel.*

249. *Principal Foci and Focal Length.* — The focus *for parallel rays* is called the *principal focus* of the lens. It may be real or virtual, and on the principal axis or on a secondary axis. The *distance from the optical centre of a lens to the principal focus* is called the *focal length* of the lens. The greater the curvature of a lens, and the greater the refractive power of the material of which it is composed, the shorter the focal length of the lens.

Fig. 175.

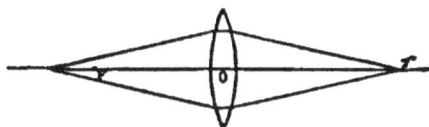

250. *Divergent Rays with Lenses.* — *Divergent* rays with a *convex* lens (Figure 175), the *point of divergence being beyond the focal length* of the lens, *become convergent* on emerging from the lens, and have a *real focus on the opposite side of the lens* to that on which the rays enter, *on the same axis as the point of divergence*, and *at a distance greater than the focal length*.

Divergent rays with a *convex* lens (Figure 176), *when the point of divergence is within the focal length* of the lens, *become less divergent* on emerging from the lens, and have a *virtual focus on the same side of the lens* as that on which the rays enter, *on the same axis as the point of divergence*, and *at a distance from the lens greater than that of the point of divergence.*

Fig. 176.

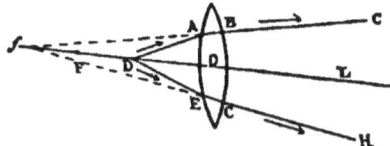

Fig. 177.

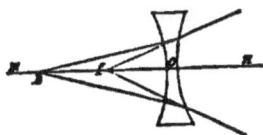

Fig. 178.

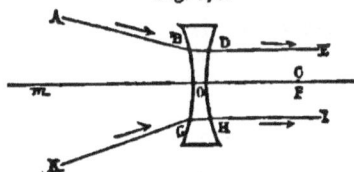

Divergent rays with a *concave* lens (Figure 177) *become more divergent*, and have a *virtual focus on the same side of the lens*

as that on which the rays enter, *on the same axis as the point of divergence,* and *nearer the lens.*

251. *Convergent Rays with Lenses.* — *Convergent* rays with a *concave* lens, the *point of convergence C* (Figure 178) being *at the focal length,* on emerging from the lens, *become parallel with the axis* on which the point of convergence lies.

When the point of convergence is *beyond the focal length* of

Fig. 179.

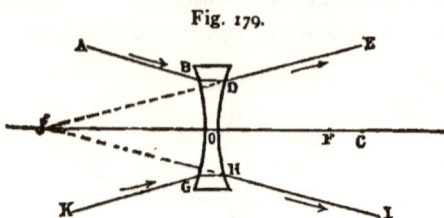

the lens (Figure 179), the rays, being less convergent on meeting the lens than in the previous case, *become divergent* on emerging from the lens, have a *virtual focus on the same side of the lens* as that on which the rays enter, *on the same axis as the point of convergence,* and *farther from the lens than the focal length* of the lens.

Fig. 180.

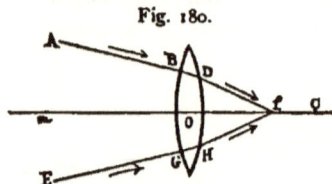

Convergent rays with a *convex* lens (Figure 180) *become more convergent* on emerging from the lens, and have a *real focus on the opposite side of the lens* to that on which the rays enter, *on the same axis as the point of convergence,* and *nearer the lens.*

252. *Images formed by Lenses.* — Rays are diverging from every point on the surface of an object; that is to say, every such point is a *point of divergence.* The *focus* of a point is a copy or *image* of that point, and *the foci of all the points on the surface of an object form an image of the object.*

To find the image of an object it is necessary to find only the foci of its extremities. To find these foci, we have only to draw axes through the extremities of the object, and locate the foci on these axes, according to the case of divergent rays under which they come.

(1.) Figure 181 represents the case of an object *A B* *beyond the focal length* of a *convex* lens. The image *a b* is *real*, because made up of real foci; *inverted*, because the axes cross between the image and the object; and in this case *larger than the object*, because *farther from the lens*.

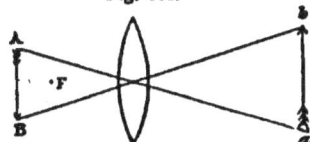

Fig. 181.

Were the object *distant*, the image would be *nearer than the object to the lens*, and consequently *smaller than the object*. The nearer the object to the principal focus of the lens, the more distant and the larger the image.

(2.) Figure 182 represents the case of an object *A B* *within the focal length* of a *convex* lens. The image *a b* is *virtual*, because made up of virtual foci; *erect*, because the axes do not cross between the image and the object; and *larger than the object*, because *farther from the lens*. The nearer the object to the principal focus of the lens, the more distant and the larger the image.

Fig. 182.   Fig. 183.

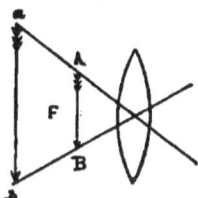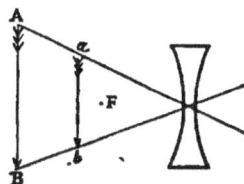

(3.) Figure 183 represents the case of an object *A B* with a *concave* lens. The image *a b* is *virtual*, because made up of virtual foci; *erect*, because the axes do not cross between the image and the object; and *smaller than the object*, because *nearer the lens*.

Virtual images can be seen only by looking through the lens at the object.

253. *Magnifying Power of Lenses.* — (1.) When an object is 40 or 50 feet distant, the rays from it which fall upon a small lens are *sensibly parallel*, and are brought to a focus

nearly at its focal length. The image of a *distant* object is, therefore, *formed nearly at the focal length* of a lens; hence *the longer the focal length the larger the image* of such an object.

(2.) When we can place the object *very near the principal focus* of the lens, *the shorter the focal length the larger the image* it will form.

Fig. 184.

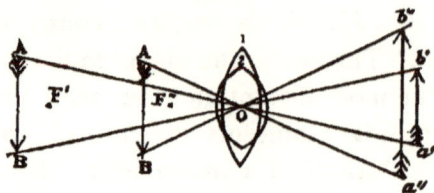

This is readily seen from Figure 184. The two lenses 1 and 2 are represented as in the same position. *F′* is the principal focus of the first lens, and *F″* that of the second lens. *A B* represents the same object placed near the principal focus of each lens, so that each will form an image of it at the same distance on the other side of the lenses. The image *a′ b′*, formed by the first lens, is seen to be smaller than the image *a″ b″*, formed by the second lens.

254. *Spherical Aberration.* — The rays which pass through an ordinary lens near the margin are *brought to a focus a little nearer the lens* than those which pass through near the centre (Figure 185). This action of the lens is called

Fig. 185.   Fig. 186.

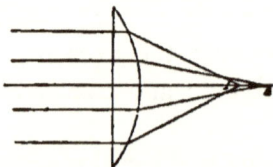
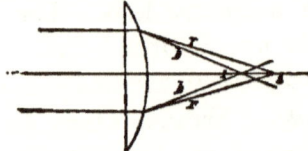

spherical aberration. It causes the image to appear blurred. It is obviated by grinding the lens to a special form, which can be exactly ascertained only by trial.

255. *Chromatic Aberration.* — An ordinary lens not only

refracts, but also *disperses* the rays of light, so that the violet rays, which are most refrangible, are brought to a focus at 1 (Figure 186), while the red rays, which are least refrangible, are brought to a focus at 2. The other rays are brought to a focus between these points. This action of the lens is called *chromatic aberration.* It causes the image to be fringed with colors. It can be overcome by combining a convex lens of crown glass with a concave lens of flint glass, which has an equal dispersive power, but a smaller refractive power. Such a combination of lenses (Figure 187) is called an *achromatic lens.*

Fig. 187.

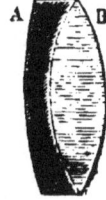

Fig. 188.

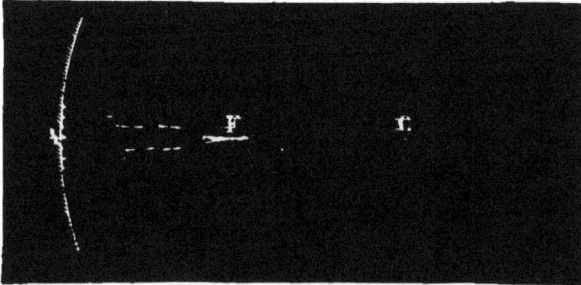

256. *Concave Mirrors correspond to Convex Lenses.* — Lenses act by *refraction*, and mirrors by *reflection*. The

Fig. 189.

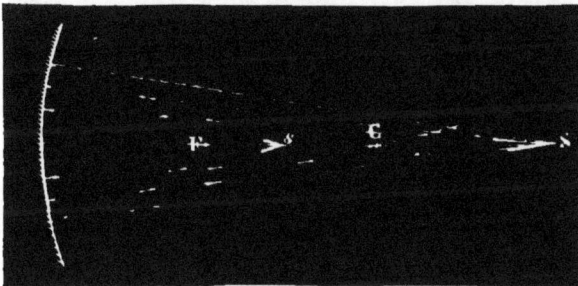

result of the action of a *concave mirror* on rays of light is *the same as that of a convex lens.* A concave mirror causes *parallel* rays after reflection to *converge to a principal focus* (Figure 188); rays *diverging from a point beyond the*

Fig. 190.

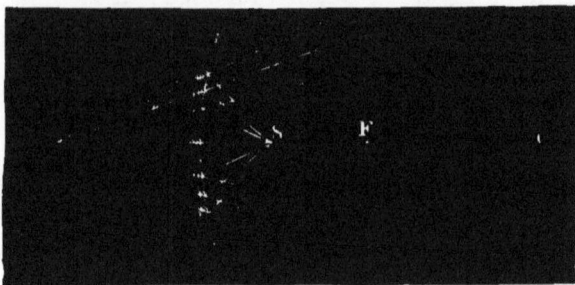

principal focus to *become convergent* (Figure 189) ; and rays *diverging from a point within the principal focus* to *become less divergent* (Figure 190).

Fig. 191.

It follows that a *concave mirror* will *form the same images as a convex lens.* The image formed by such a mirror of an object *beyond* its focal length (Figure 191) is *real and inverted;* while the image of an object placed within its focal length (Figure 192) is *virtual, erect, and larger than the object.*

To avoid spherical aberration, the reflecting surface of a concave mirror should have a curvature as nearly that of the *parabola* as possible.

Fig. 192.

The image formed by a *convex* mirror is *virtual, erect, and smaller than the object,* as in the case of a *concave* lens.

F. OPTICAL INSTRUMENTS.

257. *The Simple Microscope.* — A *simple microscope* consists of *a convex lens mounted in any convenient way.* The object is placed a little *within its focal length,* and the image seen on looking through the lens is *virtual, erect, and larger than the object.* The shorter the focal length of the lens, the greater the magnifying power of the microscope. When great magnifying power is desired, it is better to use two or more convex lenses combined so as to act as a single lens than a single lens of greater curvature.

Fig. 193.

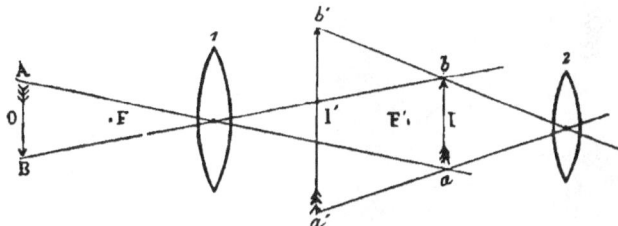

258. *The Compound Microscope and the Celestial Telescope.* — The combination of lenses employed in these two instruments is shown in Figure 193. *A B* is the object; 1 is the *objective* lens, 2 is the *eye-piece, a b* is the image formed by the objective, and *a' b'* is the image formed by

the êye-piece. The object is *beyond the focal length* of

Fig. 194.

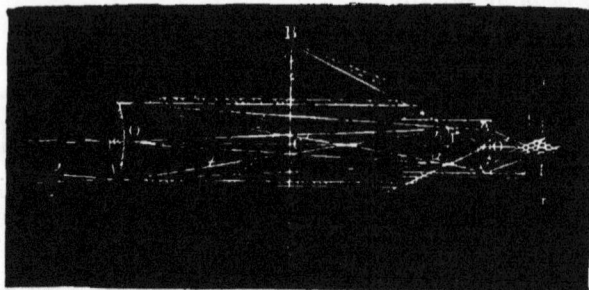

the objective. The *first image* is *real, inverted, and either larger or smaller than the object according to the distance of*

Fig. 195.

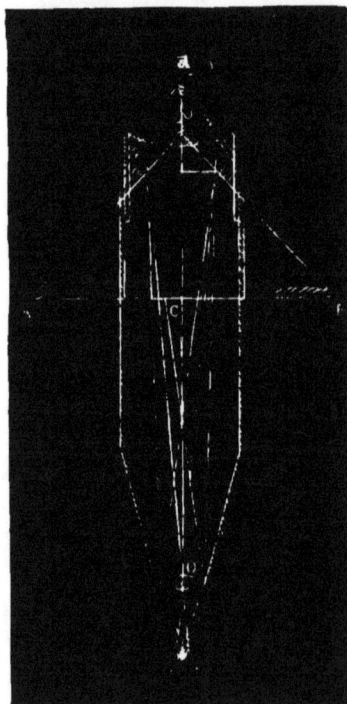

the object. The rays which meet at every point of the first image cross and diverge in front of the image as from an object. The *eye-piece* is a *simple microscope* for examining this image as if it were an object. The *image formed by the eye-piece* is *virtual, erect as compared with the first image, and larger than that image.* It is *inverted*, as compared with the object, and whether *larger or smaller than the object* depends upon *the size of the first image compared with that of the object.*

 A *telescope* is an instrument for *examining distant objects.* With the telescope the *first image* is *smaller than the object, and increases in size with the focal length of the objective.* Hence for powerful tele-scopes the objective is ground flat, so as to have as great focal length as possible, and made as large as possible, to admit the greatest possible amount of light.

The largest objectives now made are from 26 to 30 inches in diameter, with a focal length of from 30 to 40 feet. They are made achromatic. '

A *microscope* is an instrument *for examining minute objects.* The object, being under our control, can be placed as near the lens as we please, and hence the *first image* will be *larger than the object, and the less the focal length of the objective the larger the image.* Hence, for a power-ful microscope, the objective is made of as short a focal length as possible, and since it curves very rapidly it must be very small.

The objective and eye-piece of the telescope and microscope are mounted in a tube (Figures 194 and 195). The eye-piece is movable, and adjusted so that the final image is about 10 inches from the eye, the point of most distinct vision.

259. *The Terrestrial Telescope.* — In the celestial tele-scope objects are always seen inverted ; but this causes no inconvenience in observing the heavenly bodies. To make terrestrial objects appear erect, *a second objective is used*

Fig. 196.

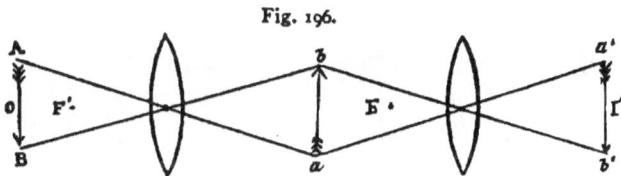

to invert the image formed by the first (Figure 196). *A B* is the object ; *a b* the image formed by the first objec-tive, which falls without the focal length of the second objective ; and *a' b'* that formed by the second objective.

260. *The Opera-Glass.* — The objective of an opera-glass is a convex lens, like that of the ordinary telescope, but the *eye-piece* is a *concave lens*. This lens is placed so that the real image of the objective would fall beyond it and outside of its principal focus (Figure 197). *A B* is

Fig. 197.

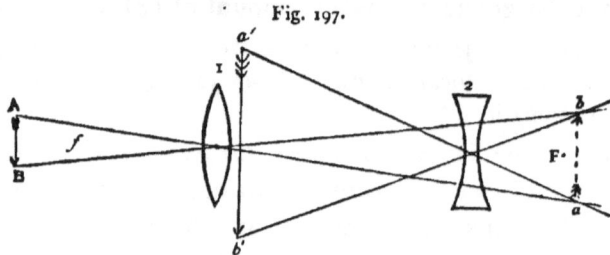

the object ; 1 the objective, 2 the eye-piece, *a b* the image that would be formed by the objective alone, and *a' b'* the image formed by the eye-piece. The rays which meet the eye-piece from the objective are converging to points between *a* and *b*. The final image *a' b'* is virtual, erect, and larger than the first image would have been. The two tubes of an opera-glass are exactly alike. They are used because it is less fatiguing to use both eyes than only one.

261. *The Reflecting Telescope.* — In reflecting telescopes the place of an object-glass is supplied by a *concave mirror*, called a *speculum*, usually composed of solid metal. The *real and inverted image* which it forms of distant objects is,

Fig. 198.

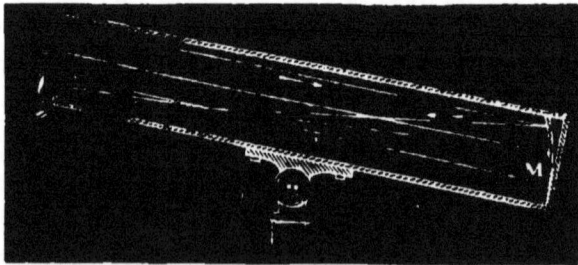

in the *Herschelian* telescope, viewed directly through an eye-piece, the back of the observer being towards the object and his face towards the speculum (Figure 198).

262. *The Camera Obscura.* — The *camera obscura* is a *dark chamber* having a movable screen within it, and a *convex lens* fitted into an opening in front. This lens forms a *real inverted image* of the objects in front, which is received upon the screen.

In order to receive the image on a horizontal table, a lens of the form shown in Figure 199 is sometimes used at the top of the camera. The rays from external objects are first refracted at the convex surface, then totally reflected at the back of the lens, which is plane, and finally emerge through the bottom of the lens, which is concave, but with a larger radius of curvature than the first surface.

Fig. 199.

The camera obscura employed by photographers (Figure 200) is a box *M N*, with a tube *A B* in front, containing an object-glass at its extremity. The object-glass is usually compound, consisting of two single lenses *E L*. At *G* is a slide of ground glass, on which the image of the scene to be depicted is thrown, in setting the instrument. The focusing is performed in the first place by sliding the part *M* of the box in the part *N*, and finally by the pinion *V*, which moves the lens. When the image has thus been rendered as sharp as possible, the sensitized plate is substituted for the ground glass.

Fig. 200.

263. *The Lantern for Projection.* — The lantern is now extensively used by teachers and lecturers for projecting experiments, diagrams, and views of various kinds upon the screen. This lantern is a kind of *reversed camera.* Some intense artificial light, as the lime light or the electric light, is enclosed in an

12

opaque box.   A convex lens, called the *condenser*, fixed in an
opening in the front of the box, condenses the light upon the
transparent picture or object to be projected.   In front of this
object is a tube containing a combination of lenses exactly like
those used with the camera.   These form a *real inverted image*
of the object on the distant screen.

264. *The Eye.* — The human eye (Figure 201) is a nearly
spherical ball, capable of turning in any direction in its socket.
Its outermost coat is thick and horny, and is opaque except in
front.   The opaque portion *H* is called the *sclerotic coat*, or the
white of the eye.   The transparent portion *A* is called the
cornea, and has the shape of a very convex watch-glass.   Behind

Fig. 201.

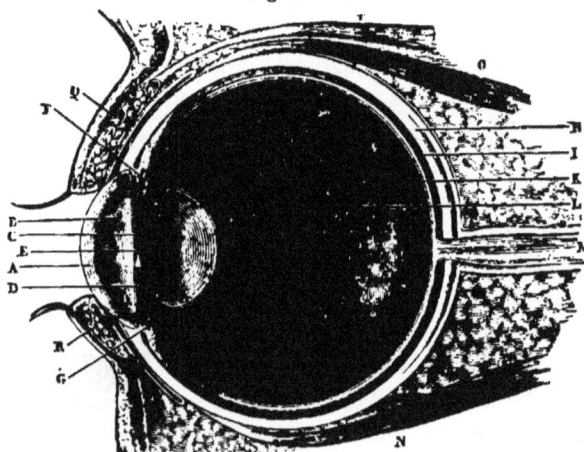

the cornea is a diaphragm *D*, called the *iris*.   It is colored and
opaque, and the circular aperture *C* in its centre is called the
pupil.   By the action of the involuntary muscles of the iris, this
aperture is enlarged or contracted on exposure to darkness or
light.   The color of the iris is what is referred to when we
speak of the color of a person's eyes.   Behind the pupil is the
crystalline lens E, which is more convex at the back than in
front.   It is built up of layers or shells, increasing in density
inwards; this tends to diminish spherical aberration.   The
cavity *B* between the cornea and the crystalline lens is called
the *anterior chamber*, and is filled with a watery liquid called

the *aqueous humor.* The much larger cavity *L,* behind the lens, is called the *posterior chamber,* and is filled with a transparent jelly called the *vitreous humor,* enclosed in a very thin transparent membrane. The posterior chamber is enclosed, except in front, by the *choroid coat I,* which is saturated with an intensely black and opaque mucus, called the *black pigment.* The choroid is lined, except in its anterior portion, with another membrane *K,* called the *retina,* which is traversed by nerve filaments diverging from the *optic nerve M.*

A pencil of rays entering the eye from an external point will undergo a series of refractions, first at the anterior surface of the cornea, and afterwards in the successive layers of the crystalline lens, all tending to render them convergent. A *real and inverted image* is thus formed of any external object to which the eye is directed. If this image falls on the retina, the object is seen; and if the image thus formed is sharp and sufficiently luminous, the object is seen distinctly.

Fig. 202.

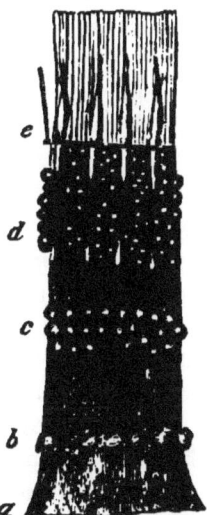

265. *The Structure of the Retina.* — Figure 202 represents a portion of the retina highly magnified, since the whole thickness of this membrane does not exceed $\frac{1}{80}$ of an inch. The inner side *a,* which is in contact with the vitreous humor, is lined with what is called the *limiting membrane.* Externally and next to the choroid coat it consists of a great number of minute rod-like and conical bodies, *e,* arranged side by side. This is the *layer of rods and cones,* and occupies a quarter of the whole thickness of the retina. From the inner ends of the rods and cones very delicate *radial fibres* spread out to the limiting membrane ; *d* and *c* are *layers of granules.* The fibres of the optic nerve are all spread out between *b* and *a.* At the entrance of the nerve these fibres predominate, and the rods and cones are wanting. At the centre of the back of the eye there is a slight circular depression of a yellowish hue, called the *macula lutea,* or yellow spot. In this spot the cones are abundant without the rods and nerve fibres.

266. *The Action of Light on the Optic Nerve.* — The distribution of the nerve fibres over the front surface of the retina would seem to indicate that they are directly acted upon by the light; but this is not the case. The fibres of the optic nerve are in themselves as blind as any other part of the body. To prove this we have only to close the left eye and with the right look steadily at the cross on this page, holding the book ten or twelve inches from the eye. The black dot will be seen quite

Fig. 203.

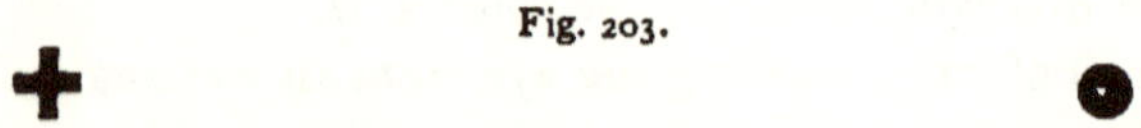

plainly as well as the cross. Now move the book slowly towards the eye, which should be kept fixed on the cross. At a certain distance the dot will suddenly disappear; but on bringing the book still nearer it will come into view again. When the dot disappears its image falls exactly upon the point where the optic nerve enters the eye, and where there are no rods and cones, but merely nerve fibres. Again, the *yellow spot* is the most sensitive part of the retina, though it contains no nerve fibres.

It would appear, then, that the fibres of the optic nerve are not directly affected by the vibrations of the ether, but only through the rods and cones.

267. *The Duration of the Impression on the Retina.* — The impression made by light on the retina does not cease the instant the light is removed, but *lasts about an eighth of a second.* If luminous impressions are separated by a less interval, they appear continuous. Thus, if a stick with a spark of fire at the end is whirled round rapidly, it gives the impression of a circle of light. The spokes of a carriage wheel in rapid motion cannot be distinguished.

The *zoetrope* illustrates the same principle. It consists (Figure 204) of a cylindrical paper box turning on an upright axis. Near the top of the box is a row of slits. The successive positions which a moving body assumes are represented in order upon a strip of paper, and this is put within the box, which is then whirled round rapidly. If we look at the figures through the slits, the successive positions come before the

Fig. 204.

eye one after another, and the impression of each lasts till the next arrives, so that they all blend into one, and the object appears to be really going through the evolutions represented.

268. *Irradiation.* -- When a white or very bright object is seen against a black ground it appears larger than it really is, while a black object on a white ground appears smaller than it

Fig. 205.

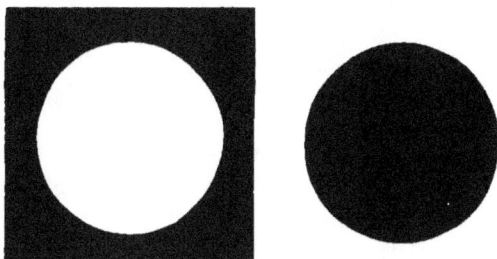

really is. The two circles in Figure 205 illustrate this. The black one and the white one have just the same diameter. This effect is called *irradiation*. It arises from the fact that the impression produced by a bright object on the retina extends beyond the outline of the image.

269. *The Optical Axis and the Visual Angle.* — A line drawn from the centre of the yellow spot through the centre of the pupil is called the *optical axis*. When we look at any object we must turn the eye so as to direct this axis towards it. This enables us to appreciate the *direction* of the object.

We have seen (252) that the image formed by a convex lens is contained between lines drawn from the extremities of the object through the centre of the lens. In the case of the crystalline lens, the angle contained between lines thus drawn is called the *visual angle* of the object, and of course measures the length of the image on the retina. All objects which have the same visual angle form images of the same length on the retina.

270. *How we estimate the Size of a Body.* — The visual angle gives us no information as to the real size of a body ; for this angle (Figure 206) diminishes as the distance of the body increases, and bodies at different distances may have the same visual angle, though they are not of the same size. Thus, *A B* and *A' B'* are

the same object ; but $A'B'$, which is farther off, has the smaller
visual angle.   Again, CD and $A'B'$ have the same visual
angle, but $A'B'$ is the larger.   We must, then, know the distance of a body in order to estimate its size ; but when we know

Fig. 206.

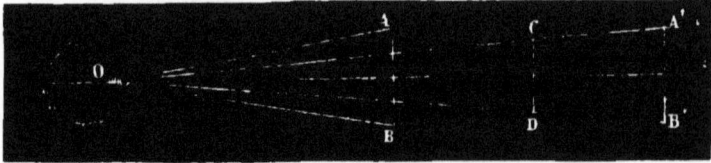

this distance, we estimate its size instinctively.   Thus, a chair
at the farthest end of the room has a visual angle only half as
large as a chair at half the distance, yet we cannot make it seem
smaller if we try.   If we are in any way deceived as to the distance of an object, we are also deceived as to its size.

271. *How we estimate the Distance of an Object.* — If we

Fig. 207.

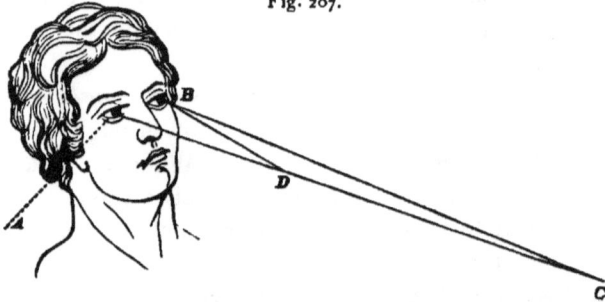

refer to Figure 207, we see that when the eyes are directed to
a distant object, as C, they are turned inward but slightly, while
they are turned inward considerably when directed to the nearer
object D.   The muscular effort we have to make in thus turning
the eyes inward so as to direct them upon an object is one of the
best methods we have of estimating its distance.

We also judge of the distance of an object from the distinct-
ness with which we see it.   The more obscure it is, the more
distant it seems.   Thus, objects seen in a fog sometimes appear
enormously large.   They appear indistinct, and we cannot rid
ourselves of the impression that they are far off ; and hence

they seem large, though they may really be small and near us.

The celebrated "Spectre of the Brocken," seen among the Hartz Mountains, illustrates the effect of indistinctness upon the apparent size of an object. On a certain ridge, just at sunrise, a gigantic figure of a man had often been seen walking, and extraordinary stories were told of him. About the year 1800 a French philosopher and a friend went to watch the spectre. For many mornings they looked for it in vain. At last, however, the monster was seen, but he was not alone. He had a companion, and, singularly enough, the pair aped all the motions and attitudes of the two observers. In fact, the spectres were merely *the shadows of the observers upon the morning fog*, which hovered over the valley between the ridges ; and because the shadows, though near, were very faint, the figures seemed to be distant, and like gigantic men walking on the opposite ridge.

When we know the real size of an object, we judge of its distance from the visual angle ; but we judge of the distance of unknown objects mainly by comparing it with the distance of known objects. This is one reason why the moon appears larger near the horizon than overhead, though she is really nearer in the latter case. When she is on the horizon, we see that she is beyond all the objects on the earth in that direction, and therefore she seems farther off than when overhead, where there are no intervening objects to help us to judge of the distance.

272. *Why Bodies near us appear Solid.* — Hold any solid object, as a book, about a foot from the eyes, and look at it first with one eye, and then with the other. It will be seen that the two images are not exactly alike. With the right eye we can see a little more of the right side of the object, and with the left eye a little more of its left side. The blending of these two pictures causes objects to appear solid.

The principle just stated explains the action of the *stereoscope.* Two photographs of an object are taken from slightly different points of view, so as to obtain pictures like those formed in the two eyes. These photographs are placed before the eyes in such a manner that each eye sees only one, but both are seen in

the same position (Figure 208). The pictures are placed at *A*
and *B;* the rays of light from them fall upon the lenses *m*
and *n*, and in passing through them are bent so that they

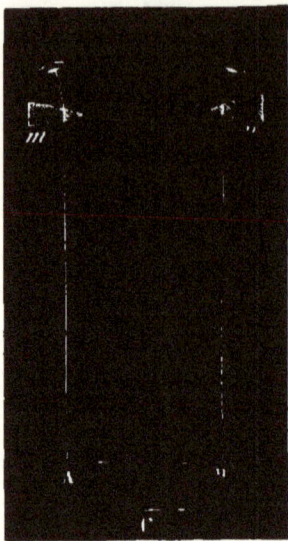

Fig. 208.

enter the eye as if they came from the
direction *C*. The lenses are portions
of a double-convex lens, arranged as
shown in the figure.

273. *Near-sighted and Far-sighted
Eyes.* — To see an object distinctly, a
clear image of it must be formed on the
retina. When an object is brought
quite near the eye, it becomes indistinct, showing that the rays are now
so divergent that the lens cannot bring
them to a focus on the retina. The
nearest point at which a distinct image
is formed upon the retina is called the
near point of vision, and the greatest
distance at which such an image is
formed is called the *far point.* In
perfectly formed eyes the near point is
about 3½ inches from the eye, and the far point is infinitely distant. The distance of the near and far points, however, is not the
same for all eyes. In some cases the near point is considerably
less than 3½ inches from the eye, while the far point is only
8 or 10 inches ; in other cases the near point is 12 inches from
the eye, and the far point infinitely distant. The former are
called *near-sighted* eyes, the latter *far-sighted* ones.

It was once thought that near-sightedness was due to the too
great convexity of the cornea or the crystalline lens, or of both,
and far-sightedness to the too slight convexity of the same ; but
their real cause lies in *the shape of the eyeball,* which in far-
sighted people is flattened, and in near-sighted people elongated,
in the direction of the axis. In Figure 209 the curve *N* shows
the form of the *normal* or perfect eye, *N'* of the *far-sighted*
eye, and *N''* of the *near-sighted* eye. The parallel rays *A* and
A are brought to a focus on the retina of the normal eye, while
only the convergent rays *A'* and *A'* are brought to a focus on
the retina of the far-sighted eye, and only the divergent rays *A''*

on the retina of the near-sighted eye. *A* ", then, is the *far* point for the near-sighted eye, since the lens has now its least convexity ; and this point must be within 18 or 20 inches, since the rays from an object at a greater distance are virtually parallel, and cannot be brought to a focus on the retina. The *near*

Fig. 209.

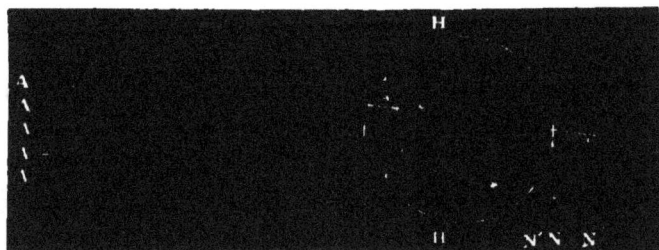

point must be less than for the normal eye, since the retina is farther from the lens, and therefore rays of greater divergence can be brought to a focus upon it. In the far-sighted eye the retina is nearer the lens than in the normal eye ; hence the near point must be farther away. While, then, the normal eye sees distant objects distinctly without adjustment, the far-sighted eye must adjust itself to see such objects.

The defect of *far-sighted* eyes can be in a great measure remedied by wearing *convex* glasses, which help to bring the rays to a focus on the retina, and thus diminish the distance of the near point. The defect of *near-sighted* eyes can be remedied by the use of *concave* glasses, which render parallel rays divergent, and thus increase the distance of the far point.

274. *Old Eyes.* — As the eye grows old it loses its power of adjusting itself for near objects, and can see distinctly only distant objects. This is quite a different thing from far-sightedness, though, like that defect, it can be remedied by the use of convex glasses.

G. Color.

275. *The Three Fundamental Qualities of Color.* — The three fundamental properties of color are *hue, purity,* and *brightness.*

The *hue* of color depends upon *the length of the luminous waves*, and corresponds to pitch in sound. In the spectrum of an incandescent solid we have all possible hues from the red through the green to the violet.

By the *purity* of a color or hue we mean its *freedom from admixture with white light*. Most natural and artificial colors contain a greater or less proportion of white light blended with their fundamental hue. This gives the hue a certain *tint* which varies with the amount of the light present. The more white light present, the paler the color.

By the *brightness* of a color we mean *the amount of light in it*. If one colored surface diffuses twice as much light as another, its color is said to be twice as bright. When a color is *at once pure and bright*, it is said to be *intense* or *saturated*.

276. *The Ideal Color-Disc.* — Every possible hue of color would be represented on a disc the color of which changed by insensible gradation from the red through the green to the violet, and then around through the purple to the red again. As such a color disc can have only an ideal existence it is called the *ideal color-disc*.

277. *The Color Chart.* — If the ideal color-disc were divided into ten equal sectors, and each sector colored with its mean hue (the hue that would result from the blending of all the hues of the sector), there would be obtained the ten following colors; *red, orange, yellow, yellowish green, green, bluish green, turquoise blue, ultramarine, violet,* and *purple*. The disc thus divided and colored (Figure 210) is called the *color chart*.

Fig. 210.

The colors of opposite sectors are *complementary* colors,

that is, colors that would *produce white when blended.* We thus obtain the five prominent pairs of complementary colors, as follows : —

> Red and bluish green,
> Orange and turquoise blue,
> Yellow and ultramarine,
> Yellowish green and violet,
> Green and purple.

If the disc had been divided into 20 equal sectors, each colored as above, there would have been produced 10 pairs of complementary colors ; if into 40 equal sectors, 20 pairs ; and so on.

278. *The Color Scale.* — In order to have the change in color equal in passing from one sector to the next in every part of the disc, it is necessary to make the sectors smaller in the region of the purple than in that of the green. In Figure 211 the disc is shown thus divided into 12 unequal sectors, each being colored as before. The colors obtained are *vermilion, orange, yellow, yellowish green, green, bluish green, turquoise blue, ultramarine, bluish violet, purplish violet, purple,* and *carmine.* This arrangement of the disc is called the *color scale.*

Fig. 211.

The colors which are nearest together on the scale form the poorest combinations, while those farthest apart form the best. When the colors are equally pure and bright, and cover equal extents of surface, the best possible combination that can be formed with any color on the scale is that formed with the *sixth* color from it. The two colors that will under similar conditions combine best with any color are the *third* colors on each side of that color on the scale.

279. *The Three Primary Colors.* — All possible hues of color can be obtained by mixing in various proportions the

three hues, *red, green,* and *violet.* Hence these three hues are called the three *primary colors.*

By mixing the hues red and green in various proportions, all the hues from red to green can be obtained. In this admixture the proportion of the red must steadily decrease and that of the green increase in passing from the red to the green. By a similar admixture of green and violet we can obtain all the hues that lie between the green and violet; and of violet and red, all the hues of purple which lie between the red and violet opposite the green. The three primary hues may be blended by means of the apparatus shown in Figure 212.

Fig. 212.

Three colored discs of thick paper are employed with it. One of the discs must be colored vermilion red, another emerald green, and the third violet. Each disc has a small hole in it at the centre, and is cut open on one side from the margin to the centre. Any two of the discs may be combined by slipping one of them into the other (Figure 213). By turning around the discs thus combined the amount of each disc exposed may be varied at pleasure.

Fig. 213.

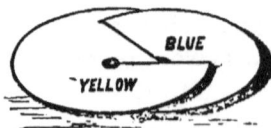

Place the red and green discs thus combined upon the rotating disc, and turn the crank rapidly. The hues of the exposed portions of the two discs will be blended in the eye (267), the impression of the color of each disc remaining on the retina till after the other disc comes round into its place. By changing the proportions of the surfaces exposed by the discs, the proportions of the two hues can be varied at pleasure. In a similar way green and violet or red and violet may be blended in various proportions.

280. *Difference between mixing Hues and mixing Pigments.* — Fill two glass tanks having parallel sides, one with a solution of aniline yellow, and the other with an ammoniacal solution of sulphate of copper, and place each in front of a lantern, so as to project two colored discs on the screen. One of these will be yellow and the other blue. Turn the lanterns till the two colored discs overlap or coincide. The resulting disc is *white* (Figure 214). In this case the hues are mixed without any mixture of substance.

Fig. 214.

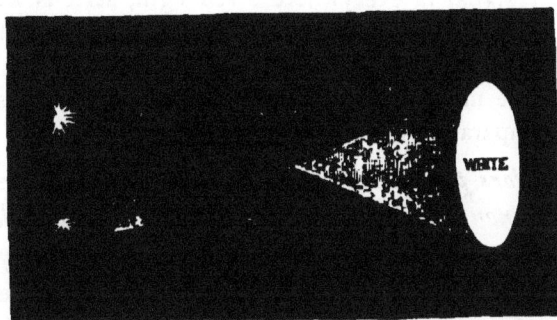

Now mix the two solutions by pouring some of each into a third cell, and place this cell before one of the lanterns. The disc on the screen will be *green*. The same result would be obtained were two cells, each containing one of the solutions, placed in front of one of the lanterns so that the light must pass through both solutions.

The yellow solution absorbs and quenches all the rays of the spectrum above the green ; and the blue solution, all those below the green. Green is the only color which is not absorbed by either substance. Hence, when light is allowed to pass through both substances, either by mixing them in one cell, or by placing them in separate cells, one in front of the other, they absorb and quench all the colors except the green, and therefore the disc obtained on the screen is green. *The hues of two colored substances are never blended when the substances themselves are mixed. One of the substances always absorbs and quenches a part of the rays which escape from the other.*

281. *Color-Blindness.* — There are many persons who cannot see certain colors. Such persons are said to be *color-blind.* Color-blindness usually takes the form of *red* blindness, though some eyes are blind to green, and others to violet.

A red-blind person can see no difference in color between a ripe strawberry and its leaf. His range of hues is limited to green and blue, and the hues produced by their combinations. Such a person will make the most absurd mistakes in attempting to match colors, mistaking a bright scarlet for a black. As the danger signal is everywhere a red light, serious accidents have occasionally been traced to color-blindness in those employed to observe the signals.

About one male in every twenty-five is more or less color-blind. Comparatively few women are color-blind.

282. *Colors produced by Absorption.* — Most of the colors of *non-luminous* bodies are produced by *absorption.* A small portion of the light that falls upon the body is diffused at the surface. The portion thus diffused enables us to see the surface, and is white or the color of the incident light. A large portion of the light is diffused from particles in the interior after it has penetrated the substance of the body to a slight depth. A portion of this light is absorbed and quenched in its passage through the substance of the body. The light which emerges from the body will be the light which enters the body minus that which has been quenched by absorption. The color of the body will be the color which is *produced by the blending of*

the hues which remain in the light after it has suffered absorption by the body. It will be *the complement of the color absorbed* by the body (277).

Bodies differ in color because they absorb different constituents of the white light that falls upon them, or else the same constituents in different proportions. In either case the hue of the light which escapes from the body will be different. A painter does not create colors. He simply prepares the surface of his canvas so that it shall destroy all the colors of white light which he does not want. He produces the hue he desires by destroying its complement. Many bodies do not have the same color by gaslight as by daylight. Some of the constituents of daylight are partially or wholly wanting in gaslight. Hence the constituents which remain after absorption are not the same in the two cases. Strictly speaking, the color does not reside in the body, but in the light which it diffuses. A body has no color in the dark.

283. *Phosphorescence and Fluorescence.* — Certain substances, after exposure to sunlight, will *appear luminous for a long time in the dark*, without any signs of combustion or of elevation of temperature. Such substances are said to be *phosphorescent.* The sulphides of calcium and of barium possess this property to a remarkable degree, and are therefore employed in the manufacture of *luminous paint.*

Fluorescence is essentially the same as phosphorescence. The former name is applied to the phenomenon observed while the body is *actually exposed to the source of light*, and the latter to the phenomenon observed *after the light from the source is cut off.* Phosphorescence is, so to speak, a kind of *persistent fluorescence.* A thick piece of uranium glass, held in the violet or ultra-violet portion of the solar spectrum, is filled to the depth of from ⅛ to ¼ of an inch with a faint nebulous light. A solution of sulphate of quinine exhibits the same effect, the luminosity in this case being bluish. If the solar spectrum is thrown upon a screen freshly washed with sulphate of quinine, the ultra-violet portion will become visible by fluorescence.

V.

MAGNETISM.

284. *Magnets.* — An iron ore was in ancient times found at Magnesia, in Asia Minor, which had a peculiar property of *attracting pieces of iron;* hence this property has been named *magnetism,* and the body possessing it is called a *magnet.* A *natural* magnet is now usually called a *lodestone.* It is one of the oxides of iron, and is very abundant in nature. *Artificial* magnets are *bars of steel,* sometimes straight and sometimes bent in the shape of a horseshoe.

285. *The Poles of a Magnet.* — If a bar-magnet is plunged into iron filings and withdrawn, the filings will

Fig. 215.

cling in large quantities to the ends of the bar and leave the middle bare (Figure 215). If the magnet is very thick in proportion to its length, the filings will adhere to all parts of it, but diminish in quantity rapidly towards the middle. The force of the magnet *resides chiefly at the ends,* which are called the *poles;* the middle line of the bar, where magnetic force is *entirely wanting,* is called the *neutral line.*

When a bar-magnet is suspended so as to turn freely, it will *take a north and south direction,* one end always turning towards the north and the other towards the south.

The former is called the *north* or *marked* pole of the magnet, and the latter the *south* or *unmarked* pole.

If the marked pole of a magnet is presented to the marked pole of another magnet which is free to turn, there is seen to be *repulsion* between the poles. The same is true if the unmarked pole of one magnet is presented to the unmarked pole of another. If we present the marked pole of one magnet to the unmarked pole of another, we see *attraction* between the poles. *Like poles of magnets repel each other, and unlike poles attract each other.*

If a magnet *A B* (Figure 216) is broken into any number of pieces, *each piece will be a complete magnet* with two poles, each

Fig. 216.

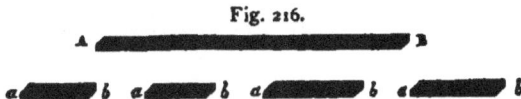

of the strength of the original poles. In each of the pieces the pole *a* to the left is the same as the pole *A* at the left end of the original magnet, and the pole *b* to the right is the same as the pole *B* of the original bar.

Fig. 217.

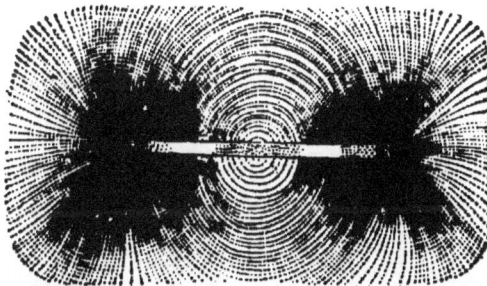

286. *Lines of Magnetic Force.* — Place a sheet of drawing-paper stretched on a frame over a bar-magnet, and sift fine iron filings upon it. Tap the paper gently, and the filings will arrange themselves in a system of curved lines (Figure 217). If a horseshoe-magnet is held under the

paper with its poles against the paper, the filings.will form
the system of curves shown in Figure 218.   These curves
mark *the lines along which the magnetic force acts*, and show
the direction and intensity of the force at each point.   The

Fig. 218.

curves are nearest together about the poles of the magnet,
where the magnetism is most intense.   The *space near a
magnet which is pervaded by its force* is called the *magnetic
field.*

287. *Magnetic Induction.* — If a bar-magnet is brought
near a piece of soft iron, it *develops magnetism* in it by an
action called *induction.*   If iron filings are scattered over
the soft iron while under the influence of the magnet, they
will adhere to its ends, as shown in Figure 219.   The soft

Fig. 219.

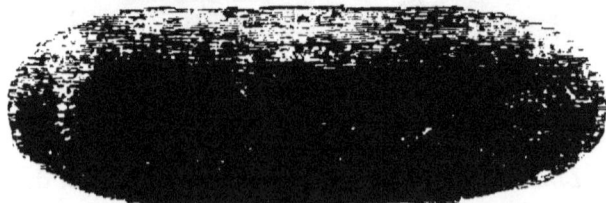

iron will have two poles and a neutral portion between
them.   The near end of the soft iron will be the opposite
pole to that of the bar presented to it ; and the far end,
the other pole.   Remove the magnet, and the iron filings

fall off from the piece of iron, showing that the iron retains no traces of magnetism, or only very slight ones.

The attraction of pieces of iron by a magnet is always preceded by induction, *the magnet developing in the portion of the iron nearest it a magnetic pole* *unlike its own.* Hence pieces of iron are attracted with equal readiness by either pole of a magnet. A piece of iron which has become magnetic by contact with a permanent magnet may attract a second piece of iron, and this a third, and so on (Figure 220). A magnetic chain may thus be formed, each component of which has two magnetic poles. Each piece of iron in the filings which cling to the poles of a magnet becomes a magnet through induction, and these pieces are held together by their dissimilar poles.

Fig. 220.

A piece of steel also becomes magnetic by induction when acted upon by a magnet, but it is not so powerfully magnetized as the soft iron. It is harder to magnetize the steel than the iron, but the steel retains its magnetic power after the magnet has been withdrawn.

288. *Magnetization of Steel Bars.* — Permanent magnets are bars of steel. These may be magnetized either by the method called *magnetization by single touch*, or by that called *magnetization by double touch*.

In the former method, the bar to be magnetized is laid on a board (Figure 221), near one end of which is a stop whose height is less than the thickness of the bar. The magnet is held in a sloping position and is drawn over the bar several times, always in the same

Fig. 221.

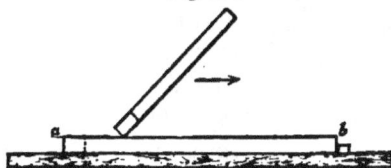

direction and with the same end down. If the marked end of the magnet is drawn over the bar from *a* to *b*, the end *a* will become a marked pole. If the magnet is drawn over the bar in the opposite direction, or the other pole of the magnet is held downward, the end *b* will become the marked pole.

In the method by *double touch*, two magnets are held one in each hand with dissimilar poles downward over the

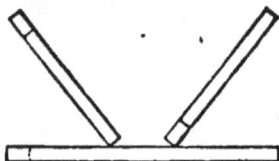

Fig. 222.

centre of the bar to be magnetized, as shown in Figure 222. They are now drawn apart quite over the ends of the bar, lifted, replaced at the centre, and again drawn over the ends. This process is repeated several times. The end of the bar over which the unmarked end of the magnet has been drawn will be the marked pole, and vice versa.

289. *Compound Magnets.* — The lifting power of a magnet generally increases with its size, but small magnets are usually able to sustain a greater multiple of their own weight than large ones. *Compound magnets* consist of a number of thin bars laid side by side, with their similar poles all pointing the same way. Figure 223 represents such a compound magnet composed of twelve elementary bars, arranged in three rows of four bars each. Their ends are inserted in masses of soft iron, the extremities of which constitute the poles of the system.

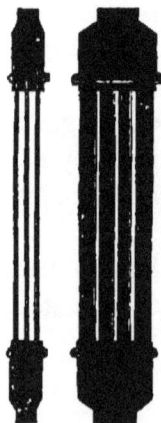

Fig. 223.

Fig. 224.

Figure 224 represents a compound horseshoe-magnet, whose poles *N* and *S* support a *keeper* of soft iron, from which is hung a bucket for holding weights. By adding fresh weights day after day, the magnet may be made to carry a greater and greater load; but if the keeper is torn away from the magnet, the additional power is instantly lost, and the magnet is able to sustain only its original load.

290. *Magnetic Needles.* — Any magnet *suspended at the centre so as to turn freely* is called a *magnetic needle.* The needle may be suspended so as to turn in a horizontal plane (Figure 225) or in a vertical plane (Figure 226). The former is called a *horizontal needle*, and the latter a *dipping needle.*

Fig. 225.   Fig. 226.

291. *Terrestrial Magnetism.* — If a steel bar, exactly balanced in a horizontal position in the frame shown in Figure 226, which is suspended by a thread, is then magnetized, it will no longer remain in equilibrium in any position in which it may be placed, but it will place itself in a particular vertical plane, and will take a particular direction in this plane. The needle takes this position in obedience to the force of *terrestrial magnetism.* The earth *acts upon the needle as if it were itself a magnet.*

The *vertical plane* of the needle is called the *magnetic meridian.* This plane usually lies several degrees from a north and south direction. The *difference between true and magnetic north,* or the *angle between the geographical*

Fig. 227.

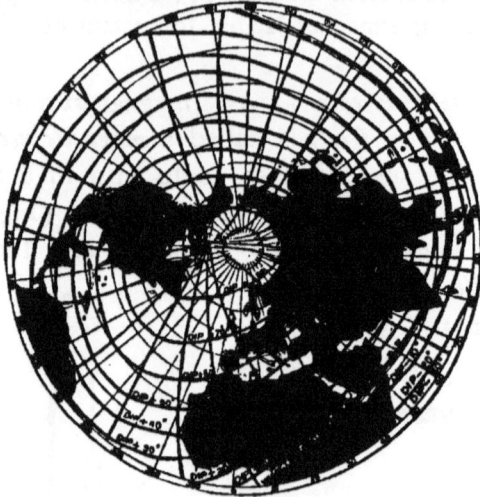

Fig. 228.

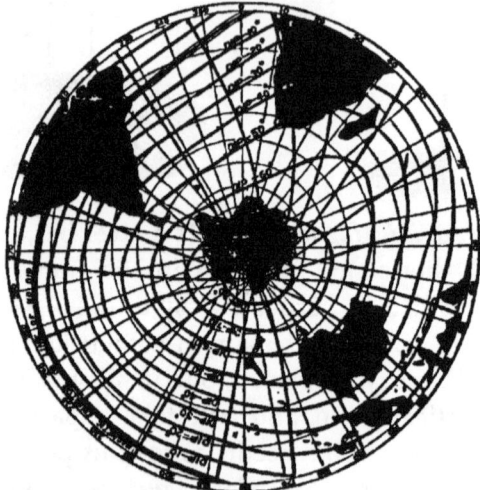

and the magnetic meridian, is called the *declination.* The direction of the needle in the vertical plane is seldom horizontal, but inclined more or less to the horizon. The angle

which the needle makes with the horizon is called the
dip. Both the declination and the dip of the magnetic
needle are very different in different parts of the earth.
As a rule, the north pole of the needle dips at places north
of the equator, and the south pole at places south of
the equator. In the neighborhood of the equator there
is *a line around the earth on which neither pole dips.* This
line is called the *magnetic equator.* The dip increases as
we proceed north and south from the magnetic equator.
The magnetic meridians and lines of equal dip are shown
in Figures 227 and 228. It will be seen that the *magnetic
poles* are at some distance from the geographic poles. The
magnetic pole north of the equator is a *south* magnetic
pole, and vice versa.

Fig. 229.

292. *The Mariner's Compass.* — The *mariner's compass* is a
declination compass *used in guiding the course of a ship.* Figure 229 represents a view of
the whole, and Figure 230 a
vertical section. It consists
of a cylindrical case, *B B'*,
which, to keep the compass

Fig. 230.

in a horizontal position in spite of the rolling of the vessel, is
supported on *gimbals.* These are two concentric rings, one

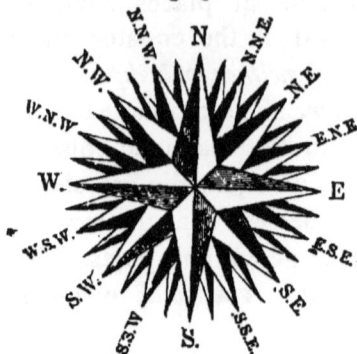

of which, attached to the case itself, moves about the axis
$x\,d$, which plays in the outer ring A; and this moves in the
supports $P\,Q$, about the axis $m\,n$, at right angles to the
first.   In the bottom of the box is a pivot, on which is placed
a magnetic bar $a\,b$, which is the
needle of the compass.   On this
is fixed a disc of mica, a little
larger in diameter than the
length of the needle, on which
is traced a star with thirty-two
branches, making the *points* of
the compass (Figure 231).   The
branch ending in a small star
(Figure 229), and marked N, is
in a line with the bar $a\,b$ (Figure
230), which is underneath the
disc.   The compass is placed near the stern of the vessel, in
sight of the helmsman.

Fig. 231.

VI.

ELECTRICITY.

I.

FRICTIONAL ELECTRICITY.

A. ELECTRICAL ATTRACTIONS AND REPULSIONS.

293. *Electrical Excitation.* — If a dry stick of sealing-wax is rubbed with a piece of dry flannel, or a vulcanite tube with a piece of dry fur, it acquires the power of attracting light bodies, such as bits of paper, pieces of straw, pith balls, etc. The body rubbed is said to be *electrified*, and the force which it manifests is called *electricity*. Electricity is *developed whenever any two unlike bodies are rubbed together*, though some bodies become electrified much more readily than others. The ancients noticed that amber, which the Greeks called *electron*, acquired the power of attracting light bodies when rubbed ; hence the terms *electrified* and *electricity*.

Electricity can be most readily and conveniently excited by rubbing a smooth vulcanite tube, 18 inches or so in length and ¾ of an inch in diameter, with a cat-skin ; or a glass tube of the same dimensions with a silk pad, composed of three or four layers of silk, and 8 or 10 inches square. The silk pad is much more effective when covered with *amalgam*, a mixture of 1 part by weight of tin, 2 parts of zinc, and 6 of mercury. The pad should be first smeared with lard, and then the powdered amalgam sprinkled over it. The tubes and rubbers work best when they are dry and hot.

294. *Electrical Attraction.* — A pith ball hung on a silk thread (Figure 232) will be attracted if we present to it either an excited glass or vulcanite tube, without allowing it to touch the ball.

Fig. 232.   Fig. 233.

An ordinary walking-stick placed in a wire loop, suspended by a narrow silk ribbon (Figure 233), may be pulled around by either of the excited tubes.

Fig. 234.

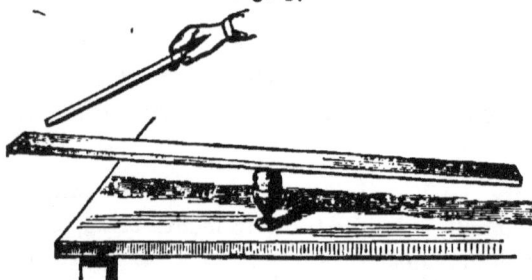

An ordinary lath balanced on an egg in an egg-cup (Figure 234) is sensibly attracted by the glass or vulcanite tube when electrified.

295. *Electrical Repulsion.* — Place an electrified glass tube in the loop shown in Figure 233, and present another excited glass tube to it. The tube in the loop will be

repelled. An electrified vulcanite tube placed in the same loop will also be repelled on presenting a second electrified vulcanite tube to it. If the pith ball of Figure 232 is allowed to touch either the electrified glass or vulcanite tube, it will soon be repelled, and it cannot again be induced to touch the tube (Figure 235).

296. *Two Kinds of Electricity.* — If an electrified vulcanite tube is placed in the wire loop of Figure 233, and an electrified glass tube is presented to it, the vulcanite will be attracted ; while, as we have seen, it will be repelled on presenting an electrified vulcanite tube to it. So, also, if an excited glass tube is placed in the loop, it will be repelled by an excited glass tube, but attracted by an excited vulcanite tube.

Fig. 235.

There are thus *two kinds of electricity :* one appearing *on glass when rubbed with silk,* and the other *on vulcanite when rubbed with fur.* The former is called *positive,* or *vitreous* electricity ; and the latter, *negative,* or *resinous* electricity.

When bodies are electrified, they are said to be *charged* with electricity. *Bodies charged with like electricities repel each other, and those charged with unlike electricities attract each other.*

297. *Electrification of the Rubber.* — The silk pad used in exciting the glass tube becomes *negatively electrified,* and the cat-skin used in exciting the vulcanite becomes *positively electrified.* Hang the vulcanite tube in the loop, having first carefully discharged the tube by rubbing the hand over it. Protect the silk pad from the hand with a piece of thin sheet-rubber. Excite the glass rod with the pad, and then present the pad to the vulcanite tube. It will be seen to attract the tube. Charge the vulcanite tube by friction with the cat-skin, and it will be repelled

by the pad which has been used in exciting a glass tube, showing that the pad is negatively electrified. Similar experiments may be tried with the cat-skin used in exciting the vulcanite tube. Whenever electricity is developed by friction, *equal quantities of both kinds of electricity are obtained, one on the body rubbed and one on the rubber.*

B. ELECTRICAL CONDUCTION AND INSULATION.

298. *Cottrell's Straw Electroscope.* — An *electroscope* is an instrument used for *indicating the presence of electricity* and also for *ascertaining whether the electricity is positive or negative.* The straw electroscope, devised by Mr. Cottrell,

Fig. 236.

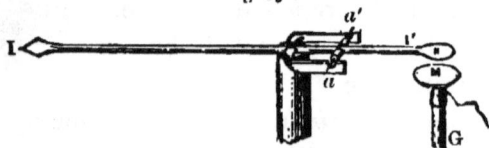

consists of a small metallic disc *M* (Figure 236), supported on a rod of glass or sealing-wax *G*, and of a smaller disc *N*, of gilt-paper, above this, fastened with sealing-wax to one end of a long straw *I I'*, capable of turning upon the needle *a a'* as an axis. The disc *N* is balanced by a little piece of bent wire at *I*, just heavy enough to separate *N* from *M*.

299. *Conductors.* — If a fine copper or iron wire is fastened to the disc *M* at one end, and coiled round the glass or vulcanite tube at the other, on exciting the tube the disc *N* is at once atracted, and the end *I* of the straw thrown upward. The attraction of the disc *N* shows that the electricity excited on the tube has passed along the wire to the disc *M*. Substances which *allow electricity to pass through them* are called *conductors* of electricity. The metals, charcoal, acids, rain-water, linen, plants, and ani-

mals are conductors. Alcohol, dry wood, paper, and straw are *semi-conductors*.

300. *Insulators.* — If the disc *M* is connected with the glass or vulcanite tube by means of a silk thread, the disc *N* will not be attracted on exciting the tube. This shows that electricity will not pass through silk. Substances *through which electricity will not pass* are called *insulators*. India-rubber, vulcanite, dry paper, hair, silk, glass, wax, sulphur, shellac, and dry air are insulators.

Conductors are said to be *insulated* when they are *completely surrounded by insulators*. A conductor may be insulated by hanging it on a silk cord or ribbon, or by supporting it on glass, vulcanite, or sealing-wax.

C. ELECTRICAL INDUCTION.

301. *Electrical Induction.* — Balance a lath upon a warm tumbler or a short rod of vulcanite (Figure 237). Place

Fig. 237.

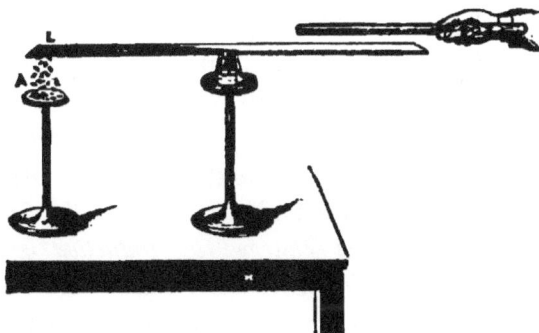

some bits of paper or elder pith upon a stand *A*, three or four inches below the end *L* of the lath, and hold an excited glass or vulcanite tube near the other end of the lath without touching it. The light bodies will be attracted, showing that the lath has been electrified. Remove the excited tube and the light bodies will fall away, showing that the lath has again become neutralized. In this case the electrification of the lath took place through the air. This *development of electricity by a charged body through an insulating medium* is called *induction*.

302. *The Electrophorus.* — The *electrophorus* consists of a plate of wax or vulcanite (Figure 238), and of a lid of

Fig. 238.

tin or brass with an insulating handle. Excite the plate by stroking it with a cat-skin, and place the lid upon it. Owing to the unevenness of the plate, the lid will touch it at comparatively few points, but the plate will act upon the lid by induction. Remove the lid, and test it with a suspended pith ball (Figure 232). It shows no signs of electrification. Replace the lid and touch it with the finger. Remove the finger and then the lid, and present the lid to the pith ball. The ball is attracted, showing that the lid is charged. Allow the pith ball to touch the lid. It is immediately repelled, having by contact become charged with the same kind of electricity as that on the lid. Present now the plate of the electrophorus to the charged pith ball, and the ball will be attracted, showing that the lid was charged with the opposite electricity to that on the plate. *Bodies charged by contact are always charged with the same electricity as that on the body acting upon them, while bodies charged by induction are always charged with the opposite electricity to that on the body acting upon them.* The lid of the electrophorus may be charged any number of times by the plate without renewing the charge on the plate.

303. *The Gold-Leaf Electroscope.* — This instrument (Figure 239) consists of two strips of gold-leaf, hung together by their upper ends to a metal rod, which passes through a hole in the top of a glass globe. The rod terminates above in a brass disc or a brass ball. The glass insulates

the disc and leaves, and protects the leaves from currents of air.

When an electrified body is placed in contact with the disc, it *charges* the disc and leaves with its own electricity, and causes the leaves to *diverge* (296).

Fig. 239.

To detect the *kind* of electricity on the charged body, first charge the leaves with a known kind of electricity, and then place the body to be tested in contact with the disc. If the leaves diverge *more* than before, the body is charged with *the same kind* of electricity as that on the leaves ; if the leaves diverge *less* than before, the body is charged with *the opposite electricity* to that on the leaves.

If a charged body is brought near the disc without touching it, the leaves will diverge, being electrified by induction. Remove the charged body, and the leaves come together again. If we present the charged body again, and touch the disc with the finger, the leaves fall together. Remove first the finger and then the charged body, and the leaves again diverge, being charged by induction with the opposite electricity to that on the charged body.

304. *Two Kinds of Electricity developed in Induction.* — *Both kinds of electricity are always developed in induction ;* the *same kind as that on the inducing body* being driven to the *far end* of the conductor, and the *opposite kind* being held on the *near end* of the conductor.

Charge the lid of the electrophorus, and hold it near one end of an insulated conductor (Figure 240). Place the *carrier* (a small metallic disc with an insulating handle) in contact

with the lid, and then with the disc of the gold-leaf electroscope, so as to charge the leaves with the electricity on the
lid. Discharge the carrier, and bring it in contact with the
far end of the insulated conductor. Again place the carrier
on the disc of the electroscope; the leaves will diverge more

Fig. 240.

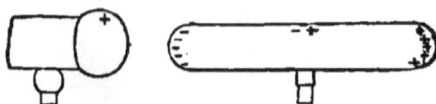

than before, showing that the far end of the conductor has upon
it the same electricity as the inducing lid. Again discharge the
carrier, and bring it in contact with the near end of the conductor. Remove the carrier again to the disc of the electroscope ; the leaves will diverge less than before, showing that
the near end of the conductor has the opposite kind of electricity
to that on the lid. In a similar way the centre of the conductor
will be found to be neutral.

While the opposite electricity to that on the inducing body
is held fast by the inducing body, the other electricity is driven
off to the farthest possible point. If two insulated conductors
are connected by a long wire, and the lid of the electrophorus
is presented to one of them, the near conductor becomes
charged with negative electricity and the far conductor with
positive electricity. If we touch a conductor under the influence
of a charged body, or connect the conductor in any way with the
earth, the far end of the conductor becomes the opposite side
of the earth, to which the electricity like that on the inducing
body is driven. Hence, when bodies connected with the earth
are acted on by induction, they have only one kind of electricity
on them, and that the opposite to that on the inducing body.
In charging a conductor by induction we must remove the earth
connection before removing the inducing body, else the electricity
which was held fast on the conductor would escape to the earth
to join the electricity which had been driven there before it.

305. *Dielectrics.* — *Induction will take place through all
insulating substances.* When an excited tube is brought
near the disc of the electroseope, the leaves diverge be-

cause of the induction which takes place through the air. If a plate of glass, of vulcanite, of paraffine, or of shellac, is held between the tube and the disc, the leaves will still diverge because of the induction which is taking place through the plate. The substance *through which induction takes place* is called a *dielectric*. *All insulators are dielectrics.*

If a metallic plate, so large that induction will not take place around it, is held between the tube and the disc of the electroscope, so as to be in connection with the earth, the leaves of the electroscope will no longer diverge. *No induction will take place through a conductor.* If the conducting plate were insulated, induction would appear to take place through it, because electricity would be developed on the far side of the plate by induction, and this electricity would carry on induction through the air.

306. *Attraction and Repulsion of Light Bodies.* — We now see why a charged body attracts a light body not previously charged. It first acts upon the light body by induction, inducing a change similar to its own on the far side of the body and an opposite change on the near side (Figure 241). The near side is attracted and the far side repelled; but the attracted side being nearer, the attraction is stronger than the repulsion, and the body as a whole is attracted. On touching the charged body it gives up to it the electricity on its near side, and so becomes charged with the same electricity as that on the charged body, and is then repelled.

Fig. 241.

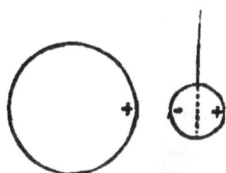

D. Electrical Potential.

307. *Electrical Potential.* — The term *potential* in Physics means *condition as regards work.* The potential of a point with respect to a force is the condition of the point as regards work done by that force. Thus, the *electrical potential* of a point is its *condition as regards work done by electricity.*

14

Electricity always *tends to move a body charged with positive electricity from a higher to a lower potential.*

When two points are at the same electrical potential, electricity does not tend to move a charged body from either point to the other, and consequently no work would be done by electricity upon a charged body in its motion from one point to the other.

In electricity, the potential of the earth is taken as zero, and the potential of a point is really *the difference between its potential and that of the earth.* Electrical potential is usually defined in terms of positive electricity. A *positive* potential is one *higher* than that of the earth, and a *negative* potential is one *lower* than that of the earth.

308. *Electrometers.* — An *electroscope* is an instrument for *detecting the presence of electricity,* and for *ascertaining its*

Fig 242.   *quality.* An *electrometer* is an instrument for *measuring the intensity of electrical attraction and repulsion,* and for *ascertaining the potential of a body.*

The *pith-ball electrometer* is shown in Figure 242. A wooden stem *C* is mounted in a metallic socket, which can be screwed to the conductor whose electrification is to be measured. A pith ball fixed to a straw stem *A* hangs from a pivot at the centre of the divided arc *B.*

Electricity is communicated from the metal socket to the ball, which is repelled. The number of degrees over which the straw passes indicates roughly the strength of the electrification of the conductor.

E. ELECTRICAL CHARGE AND DISCHARGE.

309. *The Charge entirely on the Surface.* — Suspend a tea-canister by a silk cord, and charge it as highly as possible by means of the electrophorus or other electrical machine. Lower a brass ball hung on a silk thread into

it, so as to touch the interior, and then remove it without touching the mouth of the canister. Test the ball with an electroscope, and it will be found to have brought away no electricity from the can. Bring the ball in contact with the outside of the canister and present it to the electroscope, and it will be found to have taken electricity away from the canister. By no means can any electricity be found on the inside of a hollow conductor. Hence we conclude that *the charge resides entirely on the surface;* unless, of course, a charge is developed on the inside of the hollow conductor by the induction of a charged body suspended within it.

Fig. 243.   Fig. 244.

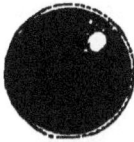
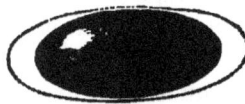

Fig. 245.   Fig. 246.

310. *Distribution of Electricity over the Surface of a Charged Body.* — Were a spherical conductor suspended on a silk thread in the centre of a large room, and charged with electricity, the charge would be distributed uniformly over the surface, as shown by the dotted line in Figure 243. The dotted lines in Figures 244, 245, and 246 show the distribution of electricity over the surface of an ellipsoid, a cylinder with rounded ends, and a disc under similar circumstances. When the conductor is *oblong*, the electricity *tends to accumulate at the ends.* The longer and thinner the conductor, the greater the accumulation at the ends.

311. *Density of the Charge.* — The *intensity of the electrification* at any point on a body is called the *electric density*

at that point. The *charge* of a body is the *quantity* of electricity on it.

The force with which electricity endeavors to escape from any portion of surface *increases with the density* at that point. The density on different parts of the surface depends upon the form of the conductor and the influence of surrounding bodies.

Charged conductors with points attached to them *become rapidly discharged by the escape of electricity from the point.* When points connected with the earth are presented to charged bodies, the bodies *become rapidly neutralized by the escape of the opposite electricity from the point to them.*

312. *Tendency of Electricity to escape from Points.* — If a sharp metallic point is fixed to one end of a small insulated conductor, and the lid of the electrophorus charged with positive electricity is held in front of the point so as to act upon the conductor by induction, negative electricity will escape from the point to the lid, and on removing the lid the conductor will be found to be charged feebly with positive electricity. If the charged lid of the electrophorus is held near the other end of the conductor, positive electricity will escape from the point, and on removing the lid the conductor will be found to be charged feebly with negative electricity. If a plate of dry glass is held between the lid of the electrophorus and the point, negative electricity will escape from the point to the glass, which will be found on examination, after removal, to be charged with negative electricity.

313. *The Electrical Machine.* — A common form of this machine is shown in Figure 247. A circular glass plate supported by a wooden frame turns between two pairs of cushions, one above and the other below the axis. In front of the plate are two metallic conductors supported on glass legs. An arm studded with metallic points directed towards the plate is connected with each of these conductors. The plate becomes charged with positive electricity by friction as it turns between the cushions, and acts upon

the points by induction. Negative electricity escapes from the points to the plate, neutralizing the positive electricity, while *positive electricity accumulates on the conductors.* The

Fig. 247.

cushions are connected with the earth to allow the negative electricity developed on them to pass off. To avoid loss of electricity from the portion of the plate which is passing from the cushions to the points, it is covered with sectors of oiled silk on both sides.

Fig. 248.

Every electrical machine may be considered as *a kind of electrical pump for raising electricity to a higher potential.* With the frictional machine only *a small quantity* of electricity is developed, but it is raised to *an enormously high potential.*

314. *The Electric Wind.* — The electricity which escapes

from a point *charges the molecules of air in front* of it, which are then *repelled* by the point. As new molecules come in to take the place of these, they are again charged and repelled. In this way *a current of air is made to set off from the point*, which may be felt by the hand or be made to flare the flame of a candle if the point is connected with the conductor of an electrical machine (Figure 248).

315. *The Electric Mill.* — The *electric mill* (Figure 249) consists of a set of metallic arms radiating horizontally from a

Fig. 249.

centre which is poised upon a point so as to turn freely. The arms are pointed at the ends and all bent around in the same direction. When the mill is connected with the conductor of an electrical machine in action, the arms revolve in a direction opposite to that in which their ends point. The motion of the mill is due to *the reaction of the molecules of the air* upon the points.

316. *The Leyden Jar.* — The *Leyden jar* consists of a wide-mouthed bottle of hard white glass (Figure 250), coated inside and out with tin-foil, except for a few inches from the mouth. The bottle is closed with a lid of hard wood, in the centre of which is a brass rod with a ball at its top. A chain hangs from the lower end of the brass rod and touches the inside tin-foil.

Fig. 250.

The inside foil can be charged with positive electricity by placing the ball near the conductor of an electrical machine, and working the machine as long as the sparks will pass. When sparks refuse to pass, the inner foil is charged almost to the potential of the conductor of the machine. *This positive charge acts inductively through the glass, and induces a negative charge on the inside of the outer tin-foil, and a positive charge on its outside.* If the outer tin foil is connected with the earth, the positive electricity is driven off into the earth, while the negative electricity is held next to the glass.

The jar may be *gradually discharged* by an arrangement shown in Figure 251. The rod connected with the inner coating has a bell upon the top of it, while a second bell on

Fig. 251.

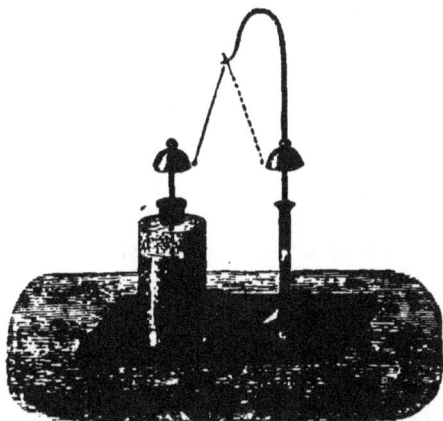

a metallic rod is connected with the outer coating by means of a strip of tin-foil on the base. A small metallic ball is hung between the bells on a silk thread. The ball is first attracted by the positive bell, and becomes charged with positive electricity. It is then repelled to the other bell, which has become negative by the release of some of the negative electricity on the outer tin-foil, owing to the removal of some of the positive electricity from the inner tin coating of the jar. It gives up its positive electricity to this bell, and is then again attracted to the positive bell.

The jar may be *suddenly discharged* by means of a *discharging rod*, as shown in Figure 252. The outside coating is touched with one end of the discharging rod, and the other end is brought near the ball, when the electricities

combine with a flash and a report. Immediately after this
has occurred, the jar is found to be completely discharged.

Fig. 252.

After a short time, however,
the jar will be found to have
acquired again a small charge.
This second charge is called
the *residual* charge.

317. *The Holtz Electrical Machine.* — This is one of the most
powerful machines ever yet invented for obtaining electricity
of high potential. In its simplest form it consists of two rather
thin discs of glass placed near together in a vertical position,
as shown in Figure 253. One of these discs is capable of
turning rapidly on a horizontal axis passing through a hole in the
centre of the other disc, which is stationary. The rotating disc
is a little smaller than the other, and has no openings in it.
There are two apertures, called *windows*, in the stationary disc
at the ends of a horizontal diameter. Just above one of these
windows and below the other, there is a paper sector fixed upon
the disc. Blunt tongues of paper run from each sector through
the window so as to touch lightly the rotating disc. In front of

Fig. 253.

the rotating disc there is a metallic comb with its points towards
the disc and just in front of the tongues from the paper sectors.
These combs are connected with the discharging rods, which

constitute the poles of the machine. Under each discharging rod is a small Leyden jar, or *condenser*.

On beginning to use the machine, it is necessary to charge the two paper sectors, one with positive and the other with negative electricity.

318. *The Spark Discharge.* — If we separate the discharging rods of a Holtz machine, and turn the disc rapidly, a torrent of sparks will pass between the rods. These sparks are due to the passage of electricity through the air between them. The spark is the *ordinary form of electrical discharge through dry gases of the ordinary density.*

The spark is *of very short duration*, lasting less than one thousandth of a second. It is very brilliant, and the impression of its light lasts much longer than the spark itself (267).

This may be shown by the following experiment. A disc (Figure 254) divided into a number of sectors alternately black and white is put into rapid rotation. The colors of the sectors blend in the eye so that they become utterly undistinguishable, and the disc appears of a uniform gray. If the whirling disc is placed in a darkened room and illuminated by a succession of electric sparks, each sector becomes perfectly distinct, and the disc appears to be standing still. The disc is visible only while the light of the spark is upon it, and the duration of the light is so short that the disc does not have time to turn an appreciable amount while illuminated by it.

Fig. 254.

The *light* of the spark is due to the fact that *the air through which the electricity passes is heated white-hot* by the electric discharge. The *sound* of the spark is due to *the sudden expansion and contraction of this heated air.*

When the spark is short it is usually straight. When it is long the spark becomes zigzag and branching, as shown in Figure 255.

319. *The Spangled Pane.* — If a number of pieces of tin-foil are arranged on a plate of glass a little way apart, and an electric discharge is allowed to pass through them, sparks will be

Fig. 255.

obtained at every interval between the pieces of foil where the electricity is obliged to pass through the air.

Very pretty effects may be obtained by pasting a long strip of tin-foil on a pane of glass in parallel lines connected at alter-

Fig. 256.

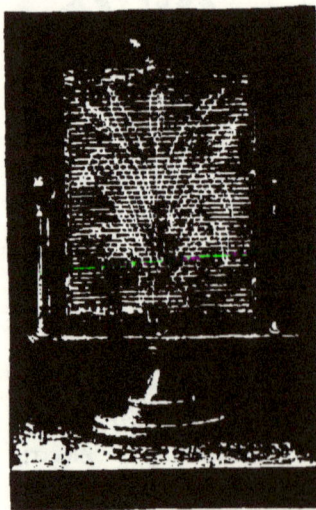

nate ends, between a knob at the top and at the bottom of the pane (Figure 256), and then tracing a design on the pane by means of a sharp point, which cuts through the strips of tin-foil wherever the lines of the pattern cross them. If a discharge is allowed to pass between the knobs, the design comes out in light, a spark being produced wherever a strip of tin-foil is cut through. Such a pane of glass is called a *spangled pane.* When the two knobs of the pane are connected with the two discharging rods of a Holtz machine in action, the effect is very pleasing. The rod or wire from one of the knobs should not quite touch the discharging rod of the machine. An interval of half an inch should be left for sparks to pass.

320. *The Auroral Discharge.* — An *auroral tube* is a long

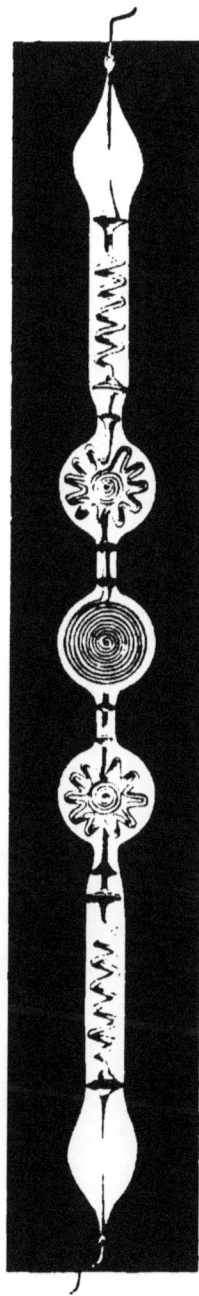

tube of glass, one or two inches in diameter, closed at the ends with brass caps through which pass metallic rods terminating within the tube and near its ends in small brass balls or points. One of the caps is fitted with a stopcock for exhaustion of the air from the interior. If this tube is screwed to the plate of an air-pump, and the caps are connected with the discharging rods of a Holtz machine, it will be found that *a longer spark can be obtained in a partial vacuum than in air of the ordinary density*. The appearance of the discharge also changes as the exhaustion proceeds. The light becomes softer and more diffused until finally the whole tube is filled with a pale luminosity. At the same time the noise of the spark is diminished till the discharge becomes inaudible.

This form of discharge, which is *common to all highly rarefied gases*, is called the *auroral discharge*, or the *vacuum discharge*. The *color* of the light *changes with the gas* used.

Tubes containing various gases in a highly rarefied state are often prepared and sealed up so as to be ready for use without the trouble of exhaustion. These tubes are called *Geissler's tubes*, or *vacuum tubes*.

The light of the auroral discharge has *great power of exciting fluorescence* (283). If any portion of the glass of the tube is colored with a fluorescent substance, as uranium, or any portion of the tube passes through a fluorescent liquid, as a solution of sulphate of quinine, when the discharge takes place, the uranium glass glows with a soft green light, and the sulphate of quinine with a soft blue, each becoming fluorescent. The accompanying plate represents a vacuum tube. The spiral portion near each end passes through a solution of sulphate of quinine contained in a wider external tube. The green portions are colored with uranium. The red shows the natural color of the discharge in rarefied air. The sulphate of quinine is quite colorless by ordinary daylight, and the uranium very nearly so.

321. *The Glow Discharge.* — When a metallic point is attached to the conductor of an electrical machine in action, it will be seen in the dark to be covered with *a soft glow of light.* A stream of molecules of air sets off from

the point (314), carrying electricity away with them, and so discharging the conductor. This discharge is called *convective discharge*. The surfaces between which convective discharge is taking place are covered with a faint glow of light. Hence convective discharge is often called *glow discharge*. In *spark* discharge the electricity *leaps from molecule to molecule through the intervening air*, while in *convective* discharge the electricity *is carried along by the molecules which traverse the intervening space*.

322. *The Brush Discharge.* — Remove the condenser from under the discharging rods of a Holtz machine, put the machine in action, and separate the rods. Instead of the ordinary spark discharge we shall find the space between the rods filled with *a pale, diffused purplish light.* From the form of this light, this discharge has been called the *brush discharge*.

Fig. 257.

The brush discharge seems to be a blending of the spark and the convective discharge. The electricity is some of the time carried by the molecules of the air, and some of the time it leaps along from molecule to molecule. In a darkened room brushes of light will be seen on various parts of a powerful Holtz machine in action. The brush sometimes assumes the form shown in Figure 257.

II.

VOLTAIC ELECTRICITY.

A. Deflection of the Needle.

323. *The Electric Current.* — The *flow of electricity through a conductor* is called the *electric current.* The phenomena of *electricity in motion,* or of current electricity, are usually classed together under the head of *voltaic electricity,* to distinguish them from those of *electricity at rest,* or of *frictional electricity.* The former department of electricity is sometimes called *dynamical electricity, electrodynamics,* or *electro-kinetics;* and the latter, *statical electricity,* or *electro-statics.*

324. *The Action of the Current on the Magnetic Needle.* — Oersted discovered, in 1819, that *a current flowing through a wire near a magnetic needle will deflect the needle.* If the

Fig. 258.

Fig 259.

wire is held *over* the needle (Figure 258), the needle will be deflected *in one direction.* If the same wire is held *under* the needle (Figure 259), the needle will be deflected *in the opposite direction.* If the current is *made to flow in the opposite direction* through the wire while over or under the needle, the needle *will be deflected in the opposite direction* to what it was before.

If *two* currents flow, one *over* the needle *in one direction,* and one *under* the needle *in the opposite direction,* they will *both tend to turn the needle the same way.*   In any case, *the stronger the current the greater the deflection* of the needle.

If the wire conveying it is *bent round the needle*, as in Figure 260, the current will flow *in opposite directions above and below* the needle.   Hence *both portions of the current will tend to turn the needle the same way*, and the deflection will be greater than when the current flowed simply over

Fig. 260.   Fig. 261.

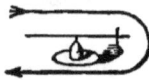
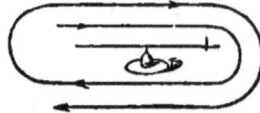

or under the needle.   If the wire is carried a second time around the needle (Figure 261), the deflection of the needle will be increased, since there will now be two currents above the needle and two below it, all tending to turn the needle the same way.

Fig. 262.

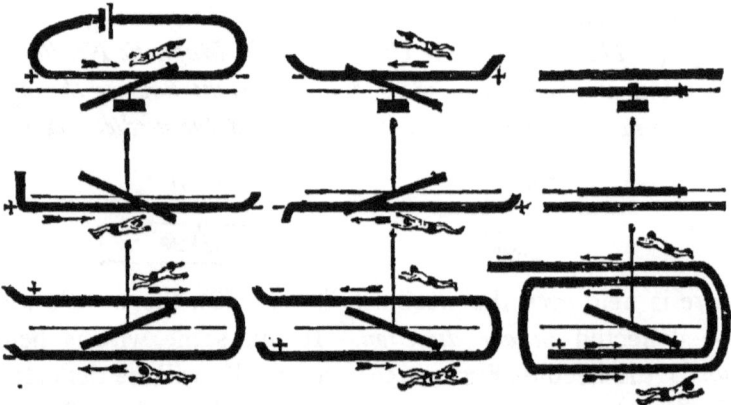

325. *Ampère's Rule.* — Ampère has given the following rule for ascertaining the direction of the deflection of the needle in any case: *Imagine a little swimmer in the electric current, always swimming with the current, and with his face to the needle.   The north pole of the needle will always be deflected to his left* (Figure 262).

326. *The Simple Galvanometer.* — A *galvanometer* is an instrument for *showing the presence, direction, and strength of an electrical current.* The *simple* galvanometer consists of *a magnetic needle,* free to turn in a horizontal or vertical plane, and *surrounded with a coil of wire.* This galvanometer shows the *presence* of a current in the wire with which it is connected, by the *deflection* of the needle ; the *direction* of the current, by the *direction of this deflection ;* and the *strength* of the current, by the *amount of the deflection.*

327. *The Astatic Needle.* — The direćtive action of the earth upon a magnetic needle impedes its deflection by the current. This action may be *neutralized* by com-

Fig. 263.

bining two needles. The needles (Figure 263) are fastened together rigidly at the centre ; and the poles of one needle are the reverse of those of the other. As there is a north and a south pole at each end, each needle must neutralize the directive action of the earth upon the other. Such a combination of needles is called an *astatic needle* (unsteady needle).

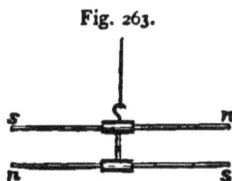

328. *The Astatic Galvanometer.* — An *astatic galvanometer* is *one in which an astatic needle is used.* The two needles of the combination are almost, but not quite, of the same strength. They are hung on a fibre of silk, and the wire is coiled around the lower needle (Figure 264). It will be seen by Ampère's rule (325) that the current that flows between the needles will tend to turn both needles the same way, while that which flows under the lower needle will tend to turn the needles in opposite directions. Owing to the greater distance, its action on the upper needle will be much feebler than its action on the lower needle. Such a galvanometer is very sensitive, since

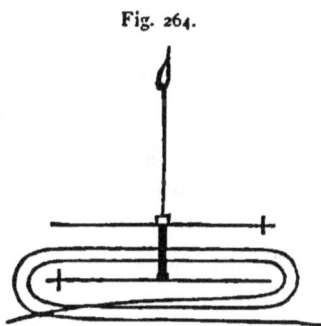

Fig. 264.

the directive action of the earth is nearly neutralized, while the effective action of the current is increased by using two needles.

When it is desired to make this galvanometer extremely sensitive, the needles are made very light, and hung on a single fibre of silk, and the wire is coiled several thousand times around the lower needle. In this case the wire is very fine, and is wound on a flat reel (Figure 265). The whole is enclosed in a glass case, to

Fig. 266.

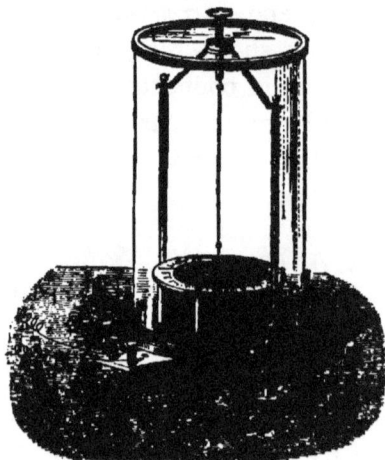

Fig. 265.

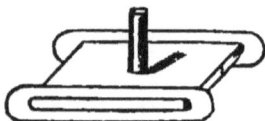

protect the needle from currents of air (Figure 266).

B. Flow of Electricity through Conductors.

329. *Electromotive Force.* — The flow of electricity through a wire connecting two conductors is analogous to the flow of water through a pipe connecting two reservoirs. When the water is at the same level in both reservoirs, no water will flow through the pipe. When the water is at different levels in the reservoirs, it will flow through the pipe from the higher level to the lower. The greater the difference between the levels, the greater the energy of the current in the pipe.

In like manner, no current of electricity will flow through a wire connecting two conductors, when the conductors are at the same potential. *When the conductors differ in potential, a current will flow through the wire from the higher potential to the lower.* The greater the difference of potential between the two conductors, the greater the energy of the current.

The *force which urges electricity through a conductor* is called the *electromotive force.* The electromotive force is *always equal to the difference of potential between the points connected by the wire.* A certain standard electromotive force has been selected as a unit, and is called a *volt.* A *conductor designed to convey a current* is called a *circuit.*

330. *Electrical Resistance.* — Every known substance offers some resistance to the passage of the current through it, but *different substances differ greatly in the amount of resistance* which they offer.

The resistance of a wire varies with its *material,* its *length,* and its *thickness. The longer and thinner* a wire, *the greater its resistance.* The metals offer comparatively little resistance to the passage of the current, and silver the least of them all. Copper stands next to silver. *The less the resistance* any substance offers to the passage of the current, *the better conductor* it is. A certain standard of resistance has been chosen as a unit, and is called an *ohm.* It is about the resistance of 250 feet of copper wire $\frac{1}{20}$ of an inch thick.

331. *The Quantity of the Current.* — By the quantity of the current we mean *the amount of electricity flowing through the circuit per second.* The *unit* of quantity is *the amount of electricity that a volt of electromotive force will cause to flow through an ohm of resistance in a second of time.* It is called a *weber.*

The power of a current to deflect a needle is directly proportional to its quantity. Hence the quantity, or volume, of the current is estimated by its power of deflecting a needle.

332. *The Division of the Current.* — When the circuit divides into two or more branches, the current will also divide among the branches in such a way that *the quantity of the current in each branch will be inversely proportional to the resistance of the branch.* Suppose the circuit divides at *A* (Figure 267) into four branches, *W, X, Y, Z,* whose resistances are in

the ratio of 3, 5, 7, and 9. Then $\frac{105}{248}$ of the current will pass through W, $\frac{63}{248}$ through X, $\frac{45}{248}$ through Y, and $\frac{35}{248}$ through Z.

Fig. 267.

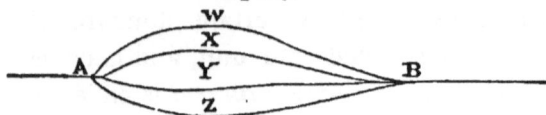

333. *The Velocity of the Current.* — The velocity of the current varies greatly under different circumstances. It ranges from about 13,000 miles a second to about 60,000 miles a second ; or from a velocity which would take it around the earth in two seconds to one which would take it twice around the earth in less than a second.

C. Electro-Chemical Action.

I. VOLTAIC BATTERIES.

334. *The Voltaic Cell.* — If two metal plates Z and C (Figure 268) are partly immersed in a liquid which acts

Fig. 268.

chemically more powerfully upon one of them than upon the other, and are placed in metallic communication outside of the liquid, either by direct contact or by means of a wire, *a current of electricity will flow outside of the liquid from the metal least acted upon by the liquid when alone to the one most acted upon.*

When two metals are thus arranged in a liquid, and are in metallic communication, the one which, if alone, would be least acted on, is entirely protected by the other. The arrangement is called a *voltaic cell*. The *portion of the plate least acted on, which is out of the liquid*, is called the *positive pole* of the cell, and the *corresponding part of the other plate* the *negative pole*.

335. *The Zinc and Copper Cell.* — In nearly all practical

forms of the voltaic cell the *negative* plate is *zinc.* The *positive* plate *varies in material.*

The simplest form of the voltaic cell consists of *a plate of copper and a plate of zinc partly immersed in dilute sulphuric acid,* which acts on the zinc, but not on the copper. With such an arrangement the current ceases after a very short time. On examination, the copper will be found to be coated with minute bubbles of hydrogen.

When a piece of zinc alone is dissolved in sulphuric acid diluted with water, it unites with the acid, forming sulphate of zinc, and sets the hydrogen of the water free. When the zinc is dissolved in the voltaic cell, sulphate of zinc is formed, but the hydrogen is liberated, not at the surface of the zinc, but at that of the copper.

The zinc of commerce, of which battery plates are made, contains many particles of iron and other metals. If a piece of ordinary zinc is placed in acid, each of these particles of iron together with the zinc near it, forms an independent small cell, and the currents produced in these small circuits cause the zinc to be rapidly consumed. The cost of chemically pure zinc prohibits its use, so a different plan is used, which is found to be in every respect equally efficacious with the employment of pure zinc. It consists in *coating the zinc with mercury.* This is done by first immersing the zinc for a few minutes in dilute sulphuric or hydrochloric acid, so as to give it a chemically clean surface, and then pouring mercury upon it. The mercury at once combines with the surface, and the zinc appears bright like silver. Zinc thus *amalgamated* is not attacked by dilute sulphuric acid, unless it forms part of a closed galvanic circuit.

336. *Two-Fluid Cells.* — In all *single-fluid* cells the compounds formed by the hydrogen in the liquid which absorbs it return to the zinc plate and retard the action on it. Cells *with two fluids* are designed to prevent this. The two principal types are *Grove's* and *Daniell's* cells. The latter is used when a constant current of moderate strength is required for days, weeks, or months ; the former, when a very powerful current is required for a few hours.

337. *Grove's Cell.* — In Grove's cell the metals used are *zinc* and *platinum;* and the fluids, *strong nitric* and *dilute sulphuric acids.* A cell of thin *porous* earthenware is filled with nitric acid, and contains the platinum plate. This cell (Figure 269) is placed within another cell of glass or vulcanite, containing the zinc and dilute sulphuric acid. The porous earthenware, when wet, permits the

Fig. 269.

Fig. 270.

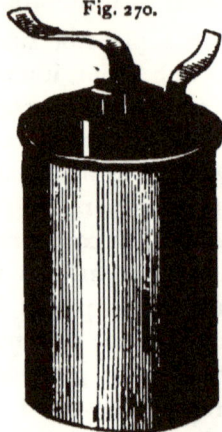

electricity to pass freely through it, while it almost entirely prevents the mixing of the liquids. The nitric acid absorbs the hydrogen as fast as it is set free.

338. *Bunsen's Cell.* — Bunsen's cell (Figure 270) is similar in construction to Grove's, with the exception that the positive plate is *carbon* instead of platinum.

Both Grove's and Bunsen's cells give off fumes of nitrous acid, which are unwholesome, and injurious to instruments. This inconvenience may be obviated by using a *solution of bichromate of potash* in the porous cup instead of nitric acid. This arrangement is the *two-fluid bichromate of potash cell.* It is much less powerful than either Grove's or Bunsen's, but is extensively used for telegraphic purposes.

339. *Daniell's Cell.* — In Daniell's cell the plates are *zinc* and *copper.* The former is usually immersed in *dilute sulphuric acid,* and the latter in a *saturated solution of sulphate of copper.*

A convenient form of this cell is shown in Figure 271. The

zinc in the form of a rod is placed inside the porous cell, which is filled with dilute sulphuric acid. The outer cell is filled with the solution of sulphate of copper. It is made of copper, and forms the positive plate of the cell.

Fig. 271.

Inside the copper cell and near the top is a copper shelf perforated with holes, on which are piled crystals of sulphate of copper. When the cell is in action, the hydrogen, as it is set free, is absorbed by the solution of the sulphate of copper which it gradually decomposes. Metallic copper is liberated from this solution and deposited upon the copper, while the zinc is gradually consumed by the sulphuric acid in the porous cup. As the solution of sulphate of copper gets weaker, a fresh portion of the sulphate is dissolved from the shelf. The power of this cell steadily decreases till the dilute acid in the porous cup is saturated with sulphate of zinc, after which it remains constant for a very long time.

340. *The Leclanché Cell.* — This consists of *zinc* and *carbon* separated by a porous cup (Figure 272). The zinc is surrounded by a *solution of sal-ammoniac,* and the carbon by a mixture of *black oxide of manganese and powdered carbon.* The cell containing the powder is filled up with water. This cell has small power, but for discontinuous work will remain in action for years, without any other attention than occasionally filling up the cell with water.

Fig. 272.

341. *The Voltaic Battery.* — The *voltaic battery* is a *combi-*

nation of voltaic cells. When the poles of a cell are not connected, they have .a certain difference of potential, which is nearly constant for each kind of cell, but varies with the different kinds of cells. When a greater difference of potential is required, it may be obtained by connecting a number of similar cells *in series,* that is, *connecting the positive pole of one cell with the negative pole of the next ;* and so on. All the poles are thus connected two by two, except in the end cells. The free positive and negative poles of these two cells are the positive and negative poles of the battery.

The difference of potential between the poles of the battery is as many times that between the poles of the cell as there are cells in the battery. In a battery of 4 cells, if we suppose the difference of potential between two poles of the same cell to be represented by the number 10, that between the poles of the battery will be represented by 40 ; if there are five cells, by 50 ; and so on.

In electrical diagrams a battery is usually represented by a series of long and thin lines and of short and thick lines. The long line at one end represents the positive pole of the battery ; and the short line at the other end, the negative pole (Figure 273).

342. *Different Ways of arranging the Battery.* — The *electromotive force* of a battery *depends solely upon the number of cells connected in series,* and *not at all upon the size of the plates.*

As the electricity has to pass through the battery as well as through the wire, the battery forms part of the circuit. Now the quantity of electricity which flows through a circuit depends upon both the electromotive force and the resistance. The greater the former and the less the latter, the greater the quantity of the current. The larger the plates of the cells the less the resistance of the battery. Hence, with the same number of cells in series, the larger the plates, the greater the quantity of the current which the battery will give.

Instead of using cells with larger plates, the cells are usually connected side by side, as shown in Figure 274. The effect of connecting cells *side by side* is not to increase the electromotive force of the battery, but to *diminish its resistance*, and so to *increase the quantity of the current.*

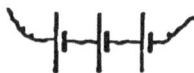

Fig. 273.

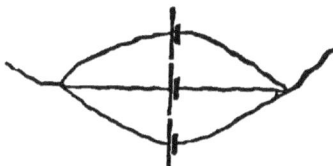

Fig. 274.

In Figure 275 twenty cells are represented as connected *in series.* Both the electromotive force and the resistance of this battery are 20 times those of a single cell of the kind employed in the battery.

Fig. 275.

In Figure 276 twenty cells are represented as connected *side by side.* The electromotive force of this battery is that of one cell only, but its resistance is only $\frac{1}{20}$ of that of one cell.

Fig. 276.

In Figure 277 twenty cells are represented as connected in a way intermediate between the last two cases. First they are arranged in series of 5 each, forming 4 compound cells, which are connected side by side. The electromotive force of this battery is 5 times that of a single cell, and its resistance is $\frac{5}{4}$ that of a single cell.

Fig. 277.

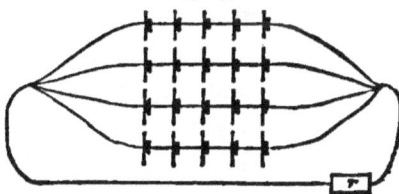

II. ELECTROLYSIS.

343. *Electrolytic Action.* — If two platinum wires, connected with the poles of a battery in action, are immersed in dilute sulphuric acid, *the acid will be decomposed.* Hydrogen will be set free at the wire connected with the negative pole of the battery, while oxygen will appear at the other wire. This can be shown to a class by placing the dilute acid in a tank with parallel glass sides, and throwing an image of the wires in the tank on a screen. Torrents of bubbles of gas will be seen to rise from the wires. The decomposition of the acid is the work of the electric current, and is called the *electrolytic action.*

If a solution of sulphate of copper is used instead of the dilute sulphuric acid, copper is deposited on the negative wire, while oxygen is set free at the positive wire.

344. *Faraday's Nomenclature of Electrolysis.* — Faraday called the decomposition of a substance by means of electricity, *electrolysis;* the substance decomposed, the *electrolyte;* the poles at which the decomposition takes place, the *electrodes;* the one connected with the positive pole of the battery the *anode,* and the one connected with the negative pole of the battery the *cathode;* the products of the decomposition, the *ions;* the one going to the anode the *anion,* and the one going to the cathode the *cation.*

345. *The Voltameter.* — The *voltameter* is an instrument for *measuring the quantity of the current.* It was invented by Faraday, and consists of a dish filled with acidulated water and fitted with electrodes (Figure 278). Receivers over the electrodes collect the gases as they are set free. The quantity of the gas liberated per minute measures the mean strength of the current during the time, and the total quantity of the gas collected measures the total quantity of electricity which has passed through the circuit.

It is necessary to collect the gases separately, as chemically clean platinum has the power to cause the hydrogen and oxy-

gen to reunite. The receiving tubes are first filled with water
and inverted over the electrodes. As the gas rises it displaces
the water. The receivers are graduated so as to show the
amount of the gas collected.

Fig. 278.

346. *Electro-Metallurgy.* — Whenever solutions of compounds of metals are decomposed, the metal is deposited
upon the cathode. This *deposition of metals by means of the
electric current* is called *electro-metallurgy*, and is of great
practical importance. The two chief processes of electro-
metallurgy are *electrotyping* and *electroplating.* The former
is *copying* by means of electricity, and the latter is *coating
the baser metals with the more noble* by means of electricity.

347. *Electrotyping.* — Anything may be electrotyped of which
a mould may be taken in wax. The chief use of electrotyping
is in *copying the face of printers' type and wood-engravings*,
after they have been arranged for the pages of a book.

A mould is first taken in wax of the article to be copied, and
the wax is coated with a thin film of some conducting substance,
such as graphite powder. The mould is then hung up in a trough
filled with a solution of sulphate of copper, called the *bath.* The
mould is connected with the negative pole of the battery, so as
to make it a cathode. A plate of copper is hung in the bath
opposite the mould, and connected with the positive pole of the
battery, so as to make it an anode. On the passage of the current through the bath, copper is deposited from the solution
upon the mould in a uniform coherent sheet, while the anode is

gradually eaten away, and keeps the bath of uniform strength. The moulds are usually hung in the bath at night, and in the morning they are removed, and the wax melted off. The copper casts are made sufficiently firm for use in printing by *backing* them with type-metal.

348. *Electroplating.* — The ordinary table-ware, such as knives, forks, tea-sets, etc., is *plated with silver by electrolysis.* The article to be plated is very carefully cleaned, and hung up in a bath containing a solution of cyanide of silver. It is then connected with the negative pole of a battery, while a piece of silver hung in front of it is connected with the positive pole. On the passage of the current, silver is deposited from the solution upon the article which forms the cathode, while the silver of the anode is gradually eaten away, and keeps the solution of uniform strength. If the article is thoroughly cleaned, and the current is maintained at the right strength, the silver will be deposited uniformly over its surface, and will adhere firmly to it.

When the article is to be *gilded*, or coated with gold, the bath must contain a solution of the cyanide of gold, and the anode must be of gold. In other respects the process is the same as in silver-plating.

In *nickel-plating* the bath contains a solution of some compound of nickel, and the anode is a piece of nickel.

D. Electro-Magnetic Induction.

349. *An Electric Whirl constitutes a Magnet.* — If a current of electricity is sent round a wire bent in the form of a

Fig. 279.

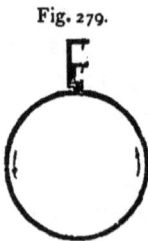

ring (Figure 279), the ring *will act in all respects like a short magnet.* The left-hand side of the ring to a person swimming round it with the current, and with his face towards the centre of the ring, will be a north pole, and the other side of the ring a south pole. If the wire is wound round and round in the form of a coil, *the multiplication of the rings will produce a stronger magnet.* By changing the strength of the current in such a coil, we change the strength of its magnetism, and *by changing the direction of the current we reverse the poles of the magnet.*

350. *The Electro-Magnet.* — If a bar of soft iron is placed within the axis of the coil, and a current sent through the coil, the iron *becomes a magnet*, with its north pole to the left hand of a person swimming around the coil with the current and with his face towards the axis of the coil. *A wire coiled round a bar of soft iron* constitutes an *electro-magnet.*

Such a magnet is *active only while the current is passing* through its coil. It loses its magnetism the moment the current stops. Its poles are reversed by reversing the current in its coil. As the strength of the current increases the magnetism of the magnet increases, but less and less rapidly, till it reaches a certain point, beyond which an increase in the strength of the current produces no increase of magnetism. At this point the magnet is said to be *saturated.* Below the point of saturation every change in the strength of the current, however slight, produces a corresponding change of magnetism.

Electro-magnets are usually made of the horseshoe form (Figure 280), and they are *much stronger than the ordinary steel magnets.* The iron core of each coil is often a separate bar, and the two bars are connected by a straight bar at the base.

Fig. 280.

351. *Magneto-Electric Currents.* — If a wire is moved in the neighborhood of a magnet in any direction whatever, except along a line of magnetic force, a difference of potential will be produced at the ends of the wire which would *cause a current to flow through a wire connecting the ends* and not acted on inductively by the magnet.

If a magnetic pole is moved in the neighborhood of a

wire, in any direction except parallel to it, *a current will be induced in the wire.* If, for instance, a magnet *N S* (Figure

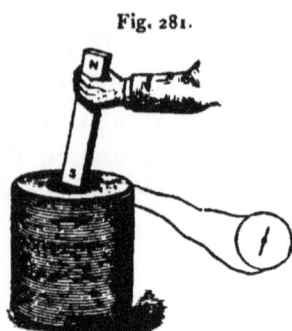

Fig. 281.

281) is moved suddenly in or out of the coil of wire, a current will be induced in the coil, which will be in one direction on inserting the pole, and in the other on withdrawing it. If the magnet is reversed so as to use the other pole, the current will be reversed.

If *a coil of wire through which a current is passing* is used instead of a steel magnet (Figure 282), precisely *similar results are obtained.* The more suddenly the steel magnet or the coil conveying a current is moved in or out of the coil, the stronger the current induced.

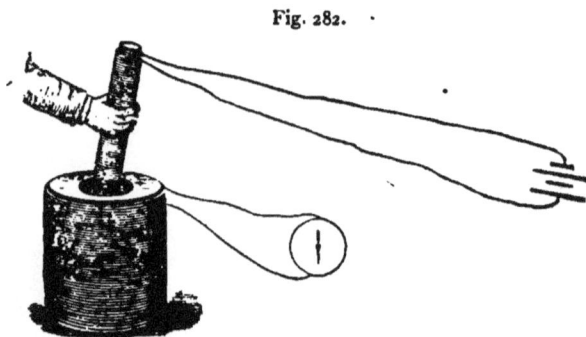

Fig. 282.

If the small coil is left within the larger coil, *any change whatever in the current in the inner coil, whether of strength or direction, will develop a current by induction in the outer coil.* So, too, if any two coils of wire, through one of which a current is passing, are near together, *any movement of the coils with respect to each other, or any change in the current in the first, will induce a current in the second.*

If a bar of *soft iron* is inserted in the inner coil of Figure 282, the current induced in the outer coil, either by motion or change of current, will be *very much stronger*.

Fig. 283.

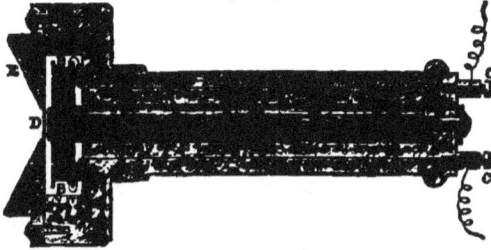

352. *The Bell Telephone.*—Figures 283 and 284 show the *Bell telephone*, in section and in perspective. It consists of a steel magnet *M* around one end of which is wound a coil of fine wire *B*. The coil and magnet are enclosed in a wooden case, which serves as a handle. One end of this case is enlarged and hollowed out at *E*, so as to serve as a mouth-piece or an ear-piece. A diaphragm of thin iron *D* is stretched across the wide end of the case, just in front of the pole of the magnet, which it does not touch.

Fig. 284.

The transmitting and receiving instruments, which are *exactly alike* in construction, are connected together by a wire. On speaking into the mouth-piece, the air in it is thrown into vibration, and the vibrations are communicated to the diaphragm. The vibrations of the iron plate produce slight temporary alterations in the magnetism of the steel magnet. These changes of magnetism in the magnet induce corresponding currents in the wire of the coil, which are transmitted over the wire which connects the two instruments. Hence pulsations of electricity exactly corresponding to the vibrations of the diaphragm of the first instrument will be transmitted over the wire and through the coil of the receiving instrument. These pulsations of the current in the coil will *induce* in the magnet of the receiving instrument *exactly the same changes of magnetism as those by which they were produced* in the sending instrument. These changes of magnetism cause the magnet to pull upon the iron plate in front of it with a varying force, and, consequently, to *make it vibrate exactly like the diaphragm of the transmitter*. These vibrations are communicated to the air, and then to the ear of the operator, which is placed at the mouth of the receiver. The words spoken into the transmitter are thus reproduced in the receiver.

Figure 285 shows the way in which the two instruments are

Fig. 285.

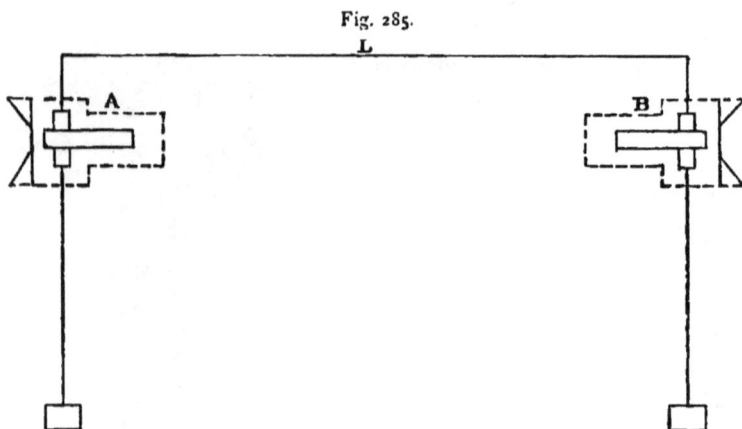

connected. The wire at each end is connected with the earth by means of a copper plate sunk in the ground, so that the circuit is completed by the earth. Otherwise two wires must be used between the instruments.

The Bell telephone is a beautiful illustration of electro-magnetic induction.

353. *The Edison Telephone.* — In the Bell telephone no battery is used. In the *Edison telephone* a battery is used, and *a current transmitted from the battery is thrown into undulations* by an arrangement called the *carbon button*. In Figure 286 *b* is a disc or button of carbon, in the form of compressed lampblack ; *a* and *c* are metallic plates placed against the front

and back of the disc. One of the poles of the battery B is connected with a, and the other with c. The current is obliged to pass from the plate a to c through the

Fig. 286.

carbon. An increase of pressure upon the metallic plates a and c diminishes the resistance of the button, either by increasing the density of the carbon or by improving the contact between the plates and the disc. The button is exceedingly sensitive to variations of pressure, the slightest alteration of pressure producing a change in the strength of the current which traverses the carbon.

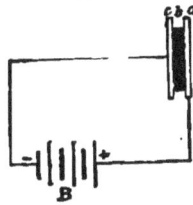

One form of the *Edison transmitter* is shown in Figure 287. The mouth-piece is of vulcanite. Back of this is the vibrating

Fig. 287.

disc, and behind this is a little hemispherical button of aluminium. This button rests upon the metallic plate in front of the carbon disc. This plate is of platinum. Behind the carbon disc is a second platinum plate, held in position by means of the screw at the back of the instrument. The battery wires are connected with the two platinum plates in such a way that the current must traverse the carbon disc.

On speaking into the mouth-piece, the disc is thrown into vibration. The vibrations are communicated to the platinum plate and the carbon disc by means of the aluminium button, thus *producing undulations in the current exactly corresponding to the vibrations of the disc.*

The *receiving instrument* of the Edison telephone is *similar to that of the Bell telephone.* Changes of magnetism are induced in it by the undulating current which traverses its coil, and these changes of magnetism *cause the disc in front of the magnet to vibrate exactly like that of the transmitter.*

354. *The Induction Coil.* — The *induction coil* consists of two coils : an inner or *primary* coil of *coarse wire,* enclosing pieces of soft iron, usually in the form of wires ; and an outer or *secondary* coil of *fine wire.* The coils are carefully insulated from each other. A current of electricity is sent through the primary coil, and *any change in the strength of this primary current develops by induction a current in the secondary coil.* The induced current is much less in quantity (331) than the primary current, but it has a far greater electromotive force (329).

355. *The Use of the Induction Coil with the Telephone.* — The induced currents from the induction coil are *better adapted for working the telephone than the direct current* from the battery. Figure 288 shows the way the coil is used with the telephone. *b* is the carbon disc of the transmitter, *a* and *c* are the platinum plates, *B* is the battery, *d* is the primary coil of the induction coil, and *e e* its secondary coil. The battery is connected with the plates *a* and *c,* and with the primary coil *d.* One end of the wire of the secondary coil is connected with the earth by the wire *G ;* and the other end to the line *L,* which runs to the receiving instrument. The undulations of the current in the primary coil induce corresponding undulations of greater electromotive force in the secondary coil. These latter undulations pass over the line, and work the receiving instrument.

Fig. 288.

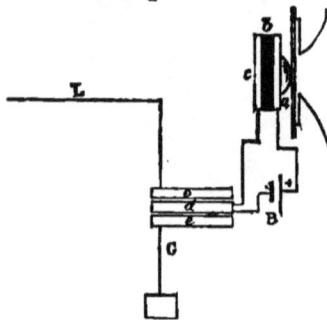

356. *The Microphone.* — When there is an imperfect contact at any point of a circuit carrying a battery current, any change

in the quality of the contact will produce a change in the strength of the current, and cause a sound in a telephone receiver included in the circuit. When the imperfect contact is between pieces of carbon lightly pressed together, *variations of the current are produced by the slightest sounds occurring near the carbons.*

· The *microphone* consists of three pieces of carbon, C, A, and C' (Figure 289). The wires from the battery B are connected with C and C' in such a way that all the pieces of carbon are in the circuit. The wires X and Y run to the receiver of a telephone. The lowest whisper spoken near the microphone is loudly reproduced in the telephone. As the carbon rod A is thrown into vibration by the pulsations of sound, it alternately lengthens and shortens. These alterations of length alternately improve and impair the contact at C and C'.

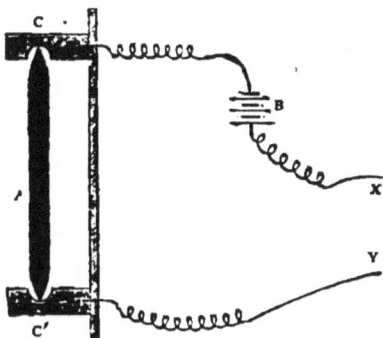

Fig. 289.

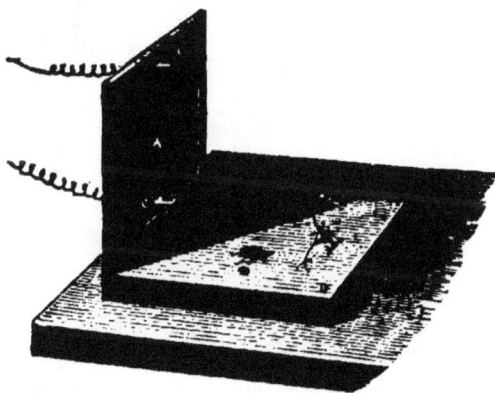

Fig. 290.

To intensify the effect, the microphone is usually placed on a sounding-board D (Figure 290). The sound caused by a fly walking on the sounding-board is distinctly audible at the distant telephone. The ticking of a watch on the sounding-board sounds like the blows of a hammer.

357. *Magneto-Electric Machines.* — The fact that electric currents are produced in a wire by any change of magnetism near it, or by moving the wire in the neighborhood of a magnet, has been utilized in the construction of machines *for the development of very powerful currents of electricity.* These machines are called *magneto-electric* or *dynamo-electric* machines. The former name is applied more especially to the machines in which the electric currents are produced *by changes of magnetism,* and the latter to those in which the currents are produced mainly *by the motion of wires in the neighborhood of magnets.* In all the dynamo-electric machines the currents are produced by revolving coils of wire between the poles of powerful horseshoe-magnets, which are sometimes steel magnets, but usually electro-magnets.

Fig. 291.

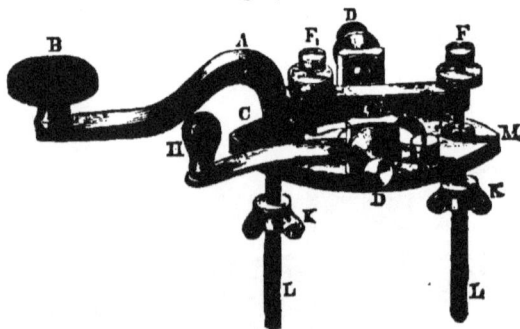

E. TELEGRAPHY.

358. *The Principal Instruments of the Simple Morse Telegraph.* — The principal instruments of the simple Morse telegraph are the *key,* the *relay,* and the *sounder.*

359. *The Key.* — The key (Figure 291) is used *for opening and closing the circuit.* Its essential parts are

shown in outline in Figure 292. *K* is the lever; *a* is the axis on which it turns; *b* is a platinum point connected with the lever; *c* is a stationary platinum point directly under *b*, called the *anvil;* and *d* is a vulcanite button by which the lever is pressed down. There is a spring under the lever of the key which keeps it up so as to separate the platinum points when the lever is not pressed down.

Fig. 292.

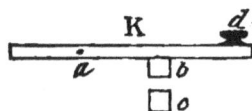

Fig. 293.

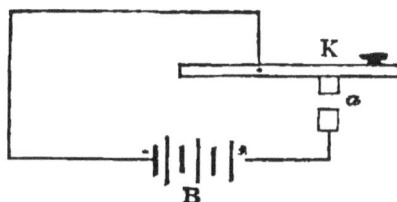

B

In Figure 293 the key is shown in the circuit of a battery. One pole of the battery is connected with the anvil by a wire, and the other with the lever at the axis. When

Fig. 294.

the lever is up, the circuit is opened at *a* by the separation of the platinum points, and the current is stopped. When the lever is pressed down, the circuit is closed by the contact of the platinum points at *a*, and the current starts.

360. *The Sounder.* — The sounder is shown in Figure
294. Its essential parts are shown in outline in Figure

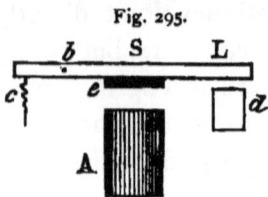

Fig. 295.

295. A is an electro-magnet; L
is a lever; b is the axis on which
the lever turns; c is a spring
which pulls the lever up; e is a
piece of soft iron, fastened across
the lever just over the electro-
magnet; and d is a piece of metal against which the lever
strikes when it is drawn down.

Figure 296 shows the sounder and key in circuit. One

Fig. 296.

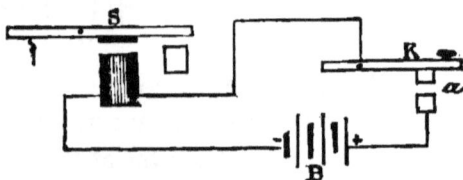

pole of the battery is connected by a wire with the circuit
of the key; the other pole is connected with one end of
the wire of the electro-magnet of the sounder, and the
other end of the wire of this magnet is connected with
the lever of the key at the axis.

When the lever of the key is up, the circuit is broken at
a, the current is stopped, the electro-magnet of the sounder
is inactive, and the lever of the sounder is thrown up by
the spring.   If the lever of the key is pushed down, contact is made at a, which closes the circuit; the current
starts, the electro-magnet of the sounder becomes active,
and the lever of the sounder is drawn down by the pull of
the magnet upon the iron above it.   As the lever is drawn
down, it *clicks* from striking the metallic stop at the end.

The clicking of the sounder is controlled by the key,
even when these are miles apart, for the sounder clicks
every time the lever of the key is depressed.   Letters and

words are indicated *by combinations of long and short intervals between the clicks.* The operator listens to the sounder just as we listen to one who is talking to us, and soon becomes able to follow it as readily.

361. *The Register.* — Sometimes an instrument called the *register* is used for receiving the message instead of the

Fig. 297.

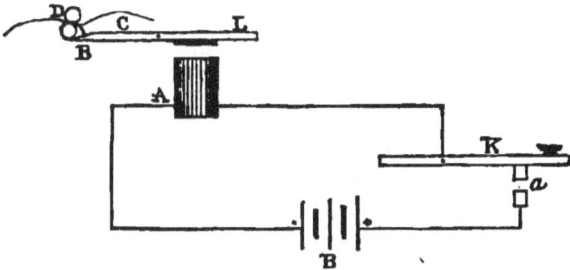

sounder. The essential parts of this instrument are shown in Figure 297. It resembles the sounder in construction and action. At the back end of the lever there is a point *B*, and just above this point a strip of paper *C* is carried

Fig. 298.

along by clockwork between two rollers at *D*. When the lever is drawn to the magnet, the point is thrown against the paper and scratches a line on it. This line will be long or short according to the time the lever is held down.

The long lines are called *dashes* and the short lines *dots.*
These dots and dashes *correspond to the short and long
intervals between the clicks of the sounder,* and their combinations form the letters of the alphabet.

362. *The Relay.* — On long lines, in which there are a number of stations, the current from the main battery is *not strong
enough to work the sounders with sufficient force.* This neces-
sitates the use of an instrument called the *relay* (Figure 298).
Its essential parts are shown in outline in Figure 299. *A* is

Fig. 299.

an electro-magnet; *l* is the lever, which
turns upon an axis at *b; c* is a piece of soft
iron fastened across the lever in front of the
electro-magnet; *f* is a spring for pulling
the lever back ; *d* and *e* are two platinum
points, the former fastened to the lever and
the latter stationary.

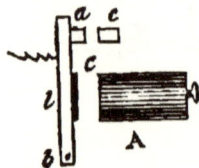

Figure 300 shows the way in which the key, relay, and
sounder are connected. The full line represents the circuit of

Fig. 300.

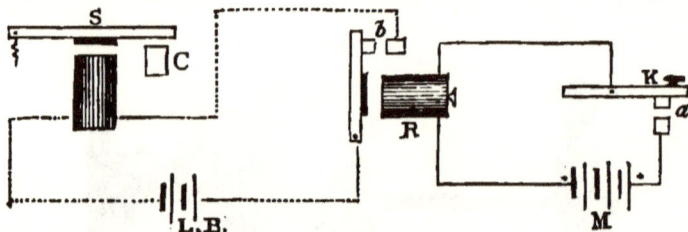

the main battery *M ;* and the dotted line, of the local battery *L.*
One pole of the main battery is connected with the anvil of the
key, and the other with one end of the wire of the electro-
magnet of the relay. The other end of the wire of this magnet
is connected with the lever of the key at the axis. One pole of
the local battery is connected to the lever of the relay, and the
other pole to the electro-magnet of the sounder and then to
the stationary platinum point of the relay. When the lever of
the key is up, the main circuit is opened at *a*, the current is
stopped, the electro-magnet of the relay is inactive, the lever
of the relay is drawn back by the spring, the local circuit is

opened at b by the separation of the platinum points, the electro-magnet of the sounder is inactive, and the bar of the sounder is thrown up by the spring. When the lever of the key is pushed down, contact is made at a, the main circuit is closed, the electro-magnet of the relay becomes active, the lever of the relay is drawn forward, contact is made at b, the local circuit is closed, the electro-magnet of the sounder becomes active, and the lever of the sounder is drawn down. Thus, *the levers of the relay and sounder vibrate in unison, but each is worked by a different battery.* The vibration of the lever of the relay is controlled by the key, and controls the vibration of the lever of the sounder by opening and closing the local circuit.

363. *The two Terminal Stations of a Line.* — Figure 301 shows the arrangement of the instruments and circuits for two terminal stations. For convenience, half of the main battery is placed at each station. There is also a key, a relay, and a sounder at each station. One pole of the main battery, say the negative, at New York is connected with the earth by a wire running to a large copper plate E sunk in the ground. A wire runs from the positive pole of the battery to the anvil of the key K, then from the lever of the key to the electro-magnet of the relay R, then from the relay to the line and along the line to Boston, then to the electro-magnet of the relay R', then to the lever of the key K', then from the anvil of the key to the negative pole of this part of the main battery, and from the positive pole of the battery to the copper plate E' in the earth. The circuit is completed by the earth, the electricity passing one way over the line and back through the earth. Each local battery is connected with its relay and sounder as in the previous section.

When the line is not in operation, the main circuit is closed at each key by pulling the side lever H seen in Figure 291 up against the anvil. This connects the axis of the lever with the anvil, and closes the circuit, although the levers of the keys are up. The electro-magnets of both relays are now active, the levers of both relays are drawn forward, both local circuits are closed, the electro-magnets of both sounders are active, and the levers of both sounders are drawn down. When the operator at one of the stations wishes to send a message, he pulls

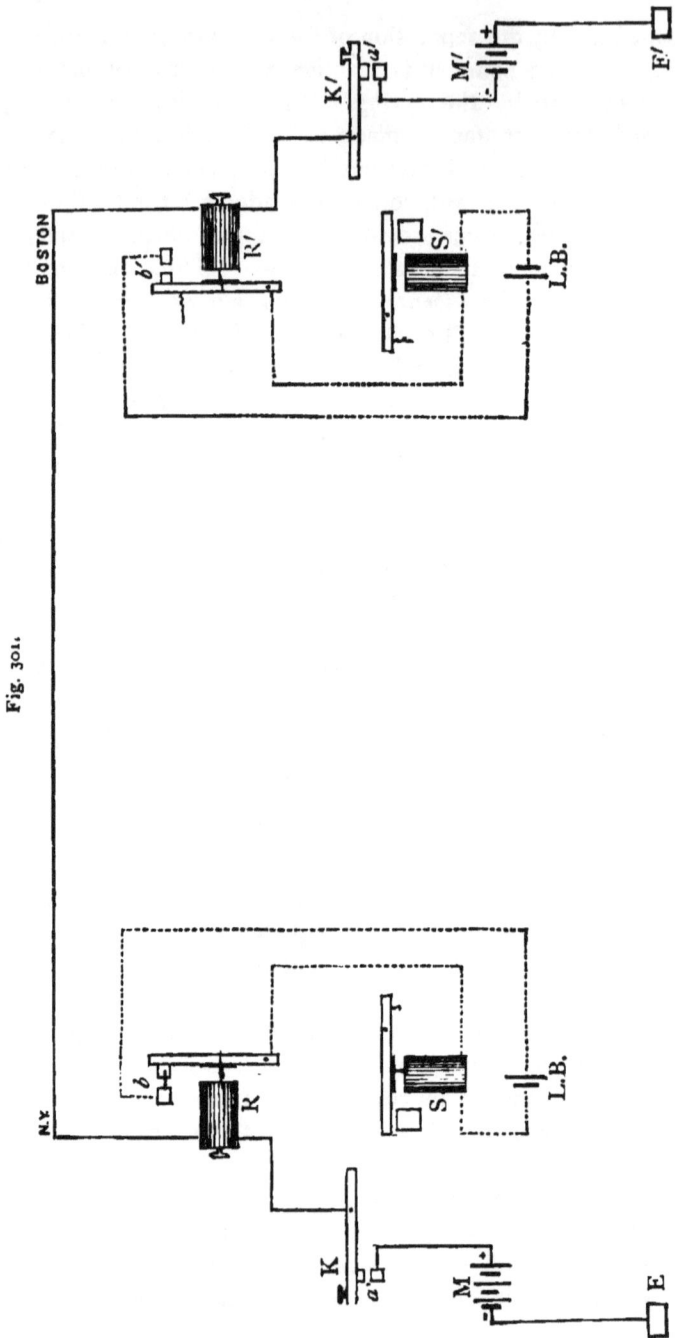

Fig. 301.

back the side lever of his key. This opens the main circuit, and causes all the electro-magnets to become inactive, and all the levers to be thrown back. On working his key, the levers of both relays and of both sounders are made to vibrate. His own sounder clicks as well as that at Boston. When the operator has finished his message, he closes his key by pulling the side lever against the anvil. Should both operators start at the same instant to send messages, the fact would be revealed by the confusion of the signals given by each sounder, and one operator would close his key and wait for the other to finish. Should the operator at the receiving station desire to interrupt the one sending the message to ask him to repeat, or for any other purpose, he has merely to open his key so as to break the circuit.

364. *A Way Station.* — One of the simplest methods of introducing the instrument of a way station into the circuit is shown in Figure 302. *A* and *B* are two brass buttons, turning on pivots at the top. Under the bottom of each button, as it stands in the diagram, is a metallic disc *D, E*. A wire runs from one of the metallic discs to the electro-magnet of the key *K'*, and thence to the anvil of the key *K'*. A wire runs from the other disc to the lever of the key. There is a third metallic disc at *C* between the buttons. When the buttons are on the discs *D* and *E*, the key and the electro-magnet of the relay are in the main circuit. The sounder and local circuit are arranged precisely as in the terminal stations. When not in operation, the key is kept closed by means of the side lever.

It will be seen at once that the levers of the relay and sounder will vibrate when the key at either terminal station is worked, and also that the levers of the relays and sounders at the terminal stations will vibrate on working the key at the way station. When the buttons *A* and *B* are both turned upon the disc *C*, the instrument of the way station will be *cut out* of the circuit, which will be completed through the buttons, these being now in contact with each other.

When any key at any station is worked, the sounders of every station which is not cut out will click. The name of the station for which the message is designed is first called, and only the operator at that station attends to the message.

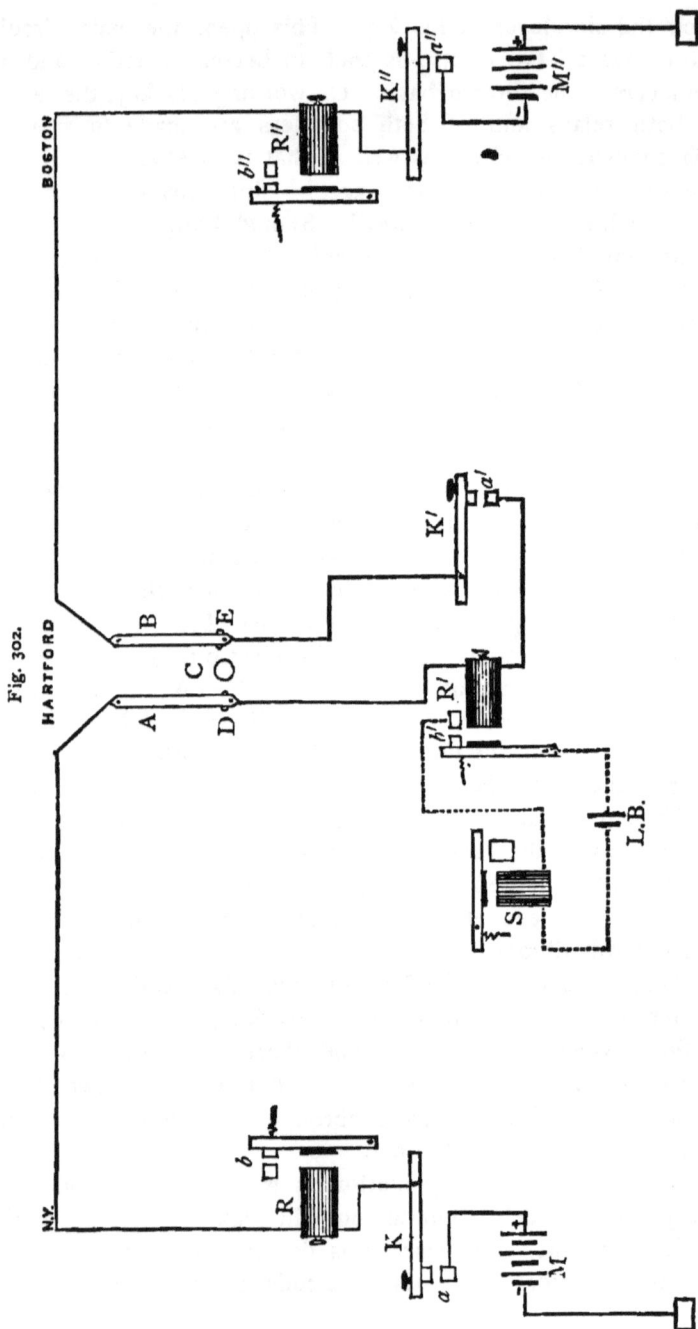

Fig. 302.

There are means at each way station to connect one of the wires with the ground and the other wire with the line on either side, so that the operator may use that side alone, in case the line is injured in any way on the other side of his station. The chief reason for dividing the main battery between the terminal stations is to enable a way station to use the line on either side in case of necessity.

F. TRANSMISSION OF POWER BY MEANS OF ELECTRICITY.

365. *Electro-Motors.* — The current produced by moving a magnet near a wire, or a wire near a magnet, always *opposes the motion which produces it;* that is to say, it *tends to produce motion in the opposite direction.* Hence, if a current of electricity from any external source were sent through the coils of a magneto-electrical machine in the direction of the one produced in these coils by the action of the machine, it would cause the cylinder to revolve in the opposite direction to that in which it must be turned to produce a current. Hence electricity when sent through the coils of such a machine becomes a source of power. A machine driven by electricity is called an *electro-motor.*

It is proposed to employ electricity as a motive power for a great variety of purposes. Companies have been formed to develop electric currents at one or more centres in cities, and send them through wires laid in the streets to the houses, to be used for a variety of domestic purposes, such as driving clocks, working sewing-machines, pumping water, etc.

It is thought that electricity will be found to be the medium by which power can be most efficiently and economically transmitted to a distance. For instance, if water-power is abundant in places remote from the localities where the power is needed, the energy of the water may be converted into that of electricity by means of dynamo-electrical machines, then the electricity conducted to the distant points through wires, and used as a source of power with similar dynamo-electrical machines (357).

G. ELECTRO-THERMAL ACTION.

366. *Thermo-Electric Piles.* — When two metals are soldered together, so as to form a closed circuit, as shown

Fig. 303.

in Figure 303, and one of the junctions is *heated* more than the other, *a current flows around the circuit.* The direction and strength of the current vary with the metals used. Such a combination of two metals is called a *thermo-electric pair.* Antimony and bismuth form the best combination among the metals. In this combination the current flows across the heated junction from the bismuth to the antimony.

With a single pair of metals only a feeble current is obtained. These pairs may be combined so as to form

Fig. 304.

batteries, or *piles.* The pairs are soldered together at alternate ends, as shown in Figure 304. Several hundred pairs are often combined in a pile.

The *least difference of temperature* between the ends of such a pile *gives rise to a current.* When used in connection with a delicate galvanometer the thermopile becomes an exceedingly sensitive *differential thermometer* (179). No current is obtained from the pile when the two faces are heated equally.

367. *The Development of Heat by means of the Current.* — Whenever a powerful current of electricity flows through a wire it *heats* it. *The finer the wire,* and *the lower the conducting power* of the material of which it is composed, *the more intense the heat* developed. *The more powerful the current* employed, *the more intense the heat with the same conductor.* Fine wires of the most refractory metals are heated white-hot, and even fused, on the passage of powerful currents.

368. *Electric Illumination by Incandescence.* — There has been for a long time an effort to make electricity available as a source of light, and at last the many practical difficulties that have been met with seem to have been nearly, if not quite, surmounted. Illumination by means of *a poor conductor heated to a white heat* on the passage of the current, is called illumination *by incandescence.* The great difficulty encountered in illumination by incandescence is that the conductor which is heated to incandescence is also apt to be destroyed by the current. Even so refractory a substance as platinum is very likely to fuse when heated to incandescence. If the current is sent through a very thin rod of carbon, the carbon becomes heated to incandescence ; but at the high temperature the carbon is liable to be destroyed by combining with the oxygen of the air. Even when the carbon is placed in an exhausted receiver, or in one which has been first exhausted of air and then filled with some gas which is a non-supporter of combustion, the rod or filament is liable to disintegration.

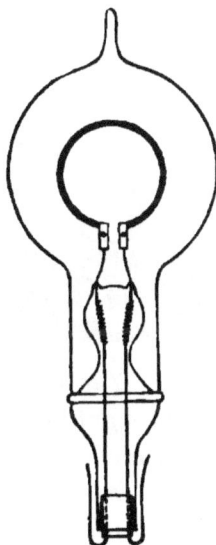

Fig. 305.

369. *The Edison Lamp.* — The Edison lamp for incandescence is shown, in section, in Figure 305. The upper portion of the lamp is a glass globe, from which the air has been exhausted, and which is hermetically sealed. In the centre of this globe is the carbon filament, bent in the form of a ring. The ends of this filament are held in little clamps, connected with the platinum wires which pass through the glass of the smaller globe under the ring, and thence out through the bottom of the lamp, where they are connected with the wires of the circuit.

The permanent success of this and similar lamps for illumination depends solely upon whether the carbon filament is found, in practice, to be sufficiently durable. The Edison filament is constructed of bamboo-wood. The resistance of the loop is from 100 to 300 ohms (330), and the amount of light that can be safely obtained from it varies from 2 to 10 candles.

These lamps will be arranged in the houses just as gas-jets are now, and electricity will be conducted to them by wires in the streets, just as gas is conducted to the gas-jets by pipes in the streets.

Edison's plan is to *measure* the electricity used in each house by a kind of voltameter (345), in which sulphate of copper is decomposed instead of sulphuric acid. The copper is deposited on one of the electrodes and so increases its weight. The increase in weight of the plate will show the amount of electricity which has passed through the instrument.

Illumination by incandescence is especially adapted for lighting rooms of the ordinary size.

Fig. 306.

370. *The Voltaic Arc.* — If two pencils of coke carbon are brought in contact in a circuit through which a powerful current of electricity is passing, and are then separated a little, *intense light and heat will be developed at the point of separation* (Figure 306). The ends of the pencils will be heated white-hot, and they will be connected by a *luminous bridge.* This bridge is called the *voltaic arc.*

The light and heat of the voltaic arc are *the most intense that can be obtained by artificial means.*

If the carbons are separated far enough to stop the current, it will not start again till they have been again brought in contact. After the current has been started, it will continue to flow after the carbons are separated, provided they are not separated too far. As the carbons

Fig. 307.

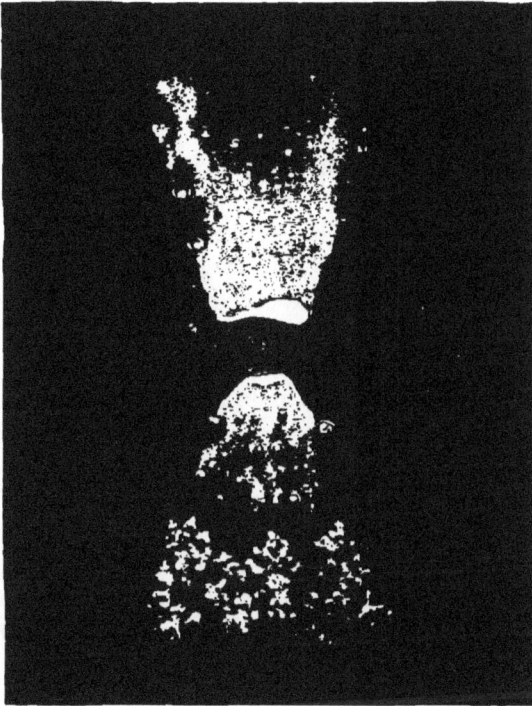

begin to separate, the current which is passing detaches little particles from each of them and transfers these to the other carbon, and so bridges over the space between the points with carbon dust. The air thus filled with particles of carbon offers less resistance to the current than the-air free from carbon dust which separates the points before they are brought into contact. Heated air moreover, offers less resistance than cold air. The intense heat of the voltaic arc is due to *the resistance which the current encounters in the space between the carbon points.*

The end of the positive carbon becomes concave, and that of the negative carbon pointed, as shown in Figure 307. Both carbons are consumed, but the positive more rapidly than the negative.

371. *Illumination by the Voltaic Arc.* — In order to obtain illumination by the voltaic arc, a lamp is needed to keep the carbons all the time *at the right distance apart,* and to bring them together, in case the current should stop, and then to separate them again when it has started.

In the best lamps for this purpose, the points are moved by means of clock-work, which is so constructed that it can be made to move the points either together or apart. The clock-work is controlled by an electro-magnet by means of a lever. The current passes through the coil of this electro-magnet on its way to the carbons. When the carbons become too far apart, the current is weakened, the lever is released, and the clock-work is made to turn so as to move the carbons together. When the carbons come too near together, the current becomes strong enough to draw the lever down, and this causes the clock-work to turn so as to separate the points. When the points are at just the right distance apart, the lever is held in such a position as to stop the clock-work entirely.

Illumination by the voltaic arc is too intense for rooms of the ordinary size, but is especially adapted for out-door illumination, and for large halls and workshops.

VII.

METEOROLOGY.

I.

CONSTITUTION OF THE ATMOSPHERE.

372. *The Term Meteorology.* — The term *meteor* was formerly applied to any natural phenomenon occurring within the limits of the atmosphere ; hence the term *meteorology* as applied to that branch of Natural Philosophy which *treats of the atmosphere.*

373. *The Composition of the Atmosphere.* — The atmosphere is composed chiefly of *oxygen and nitrogen in a state of mechanical mixture*, and not of chemical combination. In every 100 volumes of air there are nearly 79.1 volumes of nitrogen and 20.9 volumes of oxygen. Owing to the tendency of these two gases to diffuse into each other, and to the currents which exist in the atmosphere, these proportions are sensibly the same in all parts of the globe and at all accessible elevations above its surface.

In addition to the oxygen and nitrogen, the atmosphere contains also a little *carbonic acid* and *watery vapor.* The amount of carbonic acid varies, in the open country, from 4 to 6 parts in a thousand. The amount of moisture is very variable, ranging from 4 parts in one hundred to 1 part in a thousand.

374. *The Height of the Atmosphere.* — The atmosphere is held to the earth by gravity, and it must terminate at that height at which *the attraction of the earth is balanced*

by the repulsion of the particles of the air. At the height of 50 miles the atmosphere is wellnigh inappreciable in its effect upon twilight. The phenomena of lunar eclipses indicate an appreciable atmosphere to the height of 66 miles ; while the phenomena of shooting stars and of the auroral light show that such an atmosphere exists at the height of 200 or 300 miles, and probably of more than 500 miles, above the earth's surface.

375. *The Weight of the Atmosphere.* — The weight or downward pressure of the air at any point is ascertained by the use of the *barometer* (126). It is different in different parts of the earth, and is in a state of *constant fluctuation* at the same place. If we observe the height of the barometer every hour of the day, and then divide the sum of the observed heights by 24, we obtain the *mean height for the day.* By dividing the sum of the daily means for a month by the number of days in the month, we obtain the *mean height for the month.* By dividing the sum of the monthly means for a year by 12, we obtain the *mean height for the year.* If we divide the sum of the annual means for a series of years by the number of years in the period, we obtain the *mean height for the place.* This at Boston is 29.988 inches.

Fig. 308.

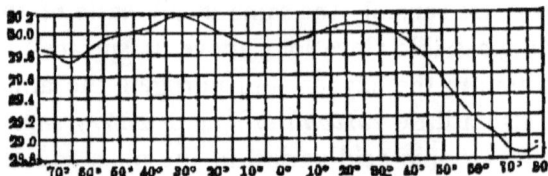

376. *The Mean Height of the Barometer at Different Latitudes.* — The curve in Figure 308 shows the mean height of the barometer at different latitudes from 75° north to 80° south. The numbers at the bottom show the latitude, and those at the side the height of the barometer in inches. The height at which the curve crosses the vertical lines of the diagram shows

the mean height of the barometer at that latitude. The height is found by following the horizontal lines to the left; and the latitude, by following the vertical lines to the bottom. It will be seen from the diagram, that the mean height of the barometer is greatest at 32° north and 25° south of the equator, and lowest at 64° north and about 70° south of the equator; also that the mean height of the barometer is *generally greater north of the equator than south of it.* There is a belt of low pressure at the equator.

377. *The Mean Height of the Barometer for Different Months.* — The mean height of the barometer varies somewhat from month to month during the year, being *generally higher in winter than in summer.* In many places the mean height in winter exceeds that of summer by half an inch, while in other places the inequality almost entirely disappears. At Pekin, China, the mean height of the barometer for January exceeds that for July by three quarters of an inch. At Boston the mean pressure does not differ more than one tenth of an inch for any two months of the year. The same is true of London and Paris. The four curves *B, L, H,* and *P* (Figure 309) show the monthly fluctuations of the mean pressure at Boston, London, Havana, and Pekin. The spaces and letters at the bottom represent the months, and the vertical lines the height.

Fig. 309.

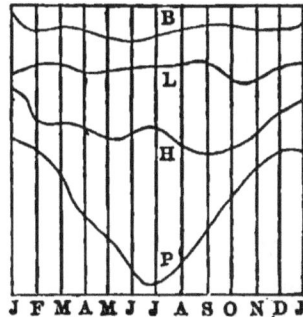

378. *Hourly Fluctuation of the Barometer.* — When the indications of the barometer for each hour of the day for a long period are averaged, it will be found that these averages are not equal. The height of the barometer is greatest about 10 A. M. and least at about 4 P. M. There are also smaller fluctuations at night, the barometer attaining a second maximum at about 10 P. M. and a second minimum at about 4 A. M. This diurnal oscillation is *greatest at the equator, and decreases as we approach either pole.*

379. *Fluctuation depending on the Position of the Moon.* — There is a small fluctuation of the barometer depending on the

position of the moon, but this variation is exceedingly minute and can be detected only by taking the mean of the most accurate observations continued for a long time. These fluctuations indicate a feeble *tide* in the atmosphere similar to those of the ocean.

380. *Irregular Fluctuations.* — The irregular fluctuations of the barometer are far greater than the periodic ones. The difference between the greatest and least heights of the barometer for a single month is called the *monthly oscillation*, and by combining observations extending over a series of years we obtain the *mean monthly oscillation*. This is *least at the equator, and increases as we proceed towards the poles.*

. At the equator it is about $\frac{1}{16}$ of an inch; in latitude 30° it is $\frac{4}{10}$ of an inch; in latitude 45°, over the Atlantic Ocean, it is 1 inch; in latitude 65° it is $1\frac{1}{3}$ inches. The extreme fluctuations are much greater than the mean monthly oscillations. The greatest and least observed heights of the barometer at Boston are 31.125 inches and 28.47 inches, the difference being 2.655 inches. The greatest observed difference at London is 3 inches; and at St. Petersburg, 3.5 inches.

II.

TEMPERATURE OF THE ATMOSPHERE.

381. *How the Atmosphere becomes Heated.* — The atmosphere becomes heated partly *by absorbing the direct rays of the sun*, partly *by contact with the warmer earth*, and partly *by absorbing the obscure heat radiated from the earth.*

A portion of the heat emitted by the sun is absorbed by our atmosphere before it can reach the earth's surface. It is estimated that on a clear day our atmosphere absorbs about one fourth of the rays which traverse it vertically. The heat thus absorbed raises the temperature of the atmosphere. It is mainly the obscure rays (220) that are absorbed by the atmosphere, and this absorption is done chiefly by the *watery vapor* in the atmosphere. The rays of the sun which reach the earth's surface are absorbed by it. The surface thus becomes heated, and communicates heat to the air which rests upon it. This heated air, becoming lighter through expansion, rises and gives place to colder air from above, which in turn becomes heated by contact with the earth.

As the surface of the earth becomes warmed by the direct rays of the sun, it radiates obscure heat back into the atmosphere. These rays are partially absorbed by the atmosphere, especially in the lower layers, where watery vapor is most abundant (224, 225).

382. *Hourly Variations of Temperature.* — The temperature of a place varies from hour to hour according to the elevation of the sun above the horizon. The average of observations taken for a long period shows that the mean hourly variations of temperature are *extremely regular.* The curve in Figure 310 shows the mean hourly variations

Fig. 310.

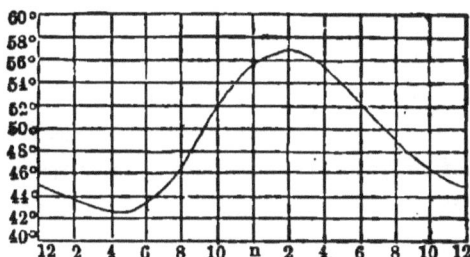

of temperature at New Haven. There is a maximum and minimum of temperature each day, the minimum occurring about an hour before sunrise, and the maximum about two hours after noon.

The highest temperature of the day, other things being equal, occurs when the amount of heat lost each instant by radiation is just equal to that received from the sun. Before midday the earth receives more heat from the sun than it loses by radiation, and the temperature rises. After noon the earth receives, each instant, less heat from the sun than it did at noon; but for some time it still receives heat faster than it parts with it. Hence the maximum of temperature occurs some time after noon. During the night we receive no direct heat from the sun, and the earth cools by radiation. About an hour before sunrise the heat received from the returning sun becomes equal to that lost by radiation, and the temperature ceases to fall.

383. *Mean Temperature of a Day.* — The mean temperature of a day is the average temperature of the 24 hours. This is found by taking the average of three observations, one at 6 A. M., one at 2 P. M., and one at 9 P. M.

384. *Monthly Variations of Temperature.* — The curves of Figure 311 show the mean temperature and also the

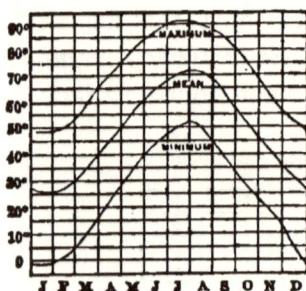
Fig. 311.

mean maximum and minimum temperatures for each month of the year at New Haven, according to observations extending through 86 years. The months are given on the horizontal line at the bottom, and the degrees of temperature on the vertical line at the left. The *warmest* months of the year for this place are *July and August*, the maximum occurring about the 24th of July. The *coldest* month is *January*, the minimum occurring about the 21st of this month. The difference between the maximum and minimum temperature is greater for the cold than for the warm months.

The chief reasons why it is colder during the winter months than during the summer months are that the sun is farther from the zenith and is a shorter time above the horizon. The earth is receiving the most heat from the sun at the time of the summer solstice, but the temperature continues to rise as long as the earth receives more heat from the sun during the day than it loses by radiation during the night. During the autumn the loss at night is much greater than the gain by day, and the temperature rapidly falls. The temperature continues to fall till the gain by day is again equal to the loss by night. This does not occur till some time after the winter solstice.

385. *Irregular Fluctuations of Temperature.* — Besides

the periodic variations of temperature, there are irregular fluctuations of temperature which are liable to occur any hour of the day and any day of the year.

386. *Variations of Temperature with the Latitude.* — As we proceed from the equator to the poles, the temperature generally falls, but not at a uniform rate, and the rate of fall is different on different meridians. Hence the lines of equal temperature on the surface of the earth do not coincide with the parallels of latitude. Lines which *connect places of equal mean temperature* are called *isothermal lines.* The isothermal lines for every ten degrees are shown on the accompanying map (Figure 312). They are much more irregular on and around the continents than in the oceans.

387. *The Temperature of the two Sides of the Atlantic.* — It will be seen from the map in Figure 312 that the mean temperature of the *eastern* side of the Northern Atlantic Ocean is *considerably higher than that of the western side* at the same latitude. The temperature of Dublin is as high as that of New York, though the former is 13° farther north, while near Lake Superior, in latitude 50°, we find the same mean temperature as at the North Cape, in latitude 72°.

The high temperature of the European coast is due to *the high temperature of the Northern Atlantic* and *the prevalent westerly winds.* The Gulf Stream conveys the warm water of the equatorial region into the North Atlantic. The temperature of the North Atlantic is thus raised considerably above what is due to its latitude, and the prevalent westerly winds of the middle latitudes carry this heat to the eastern side of the Atlantic and away from its western side.

388. *The Temperature of the two Sides of the Pacific.* — Owing to the currents of the Pacific Ocean, there is a corresponding difference of temperature between its eastern and western coast, the temperature of the *east* coast being

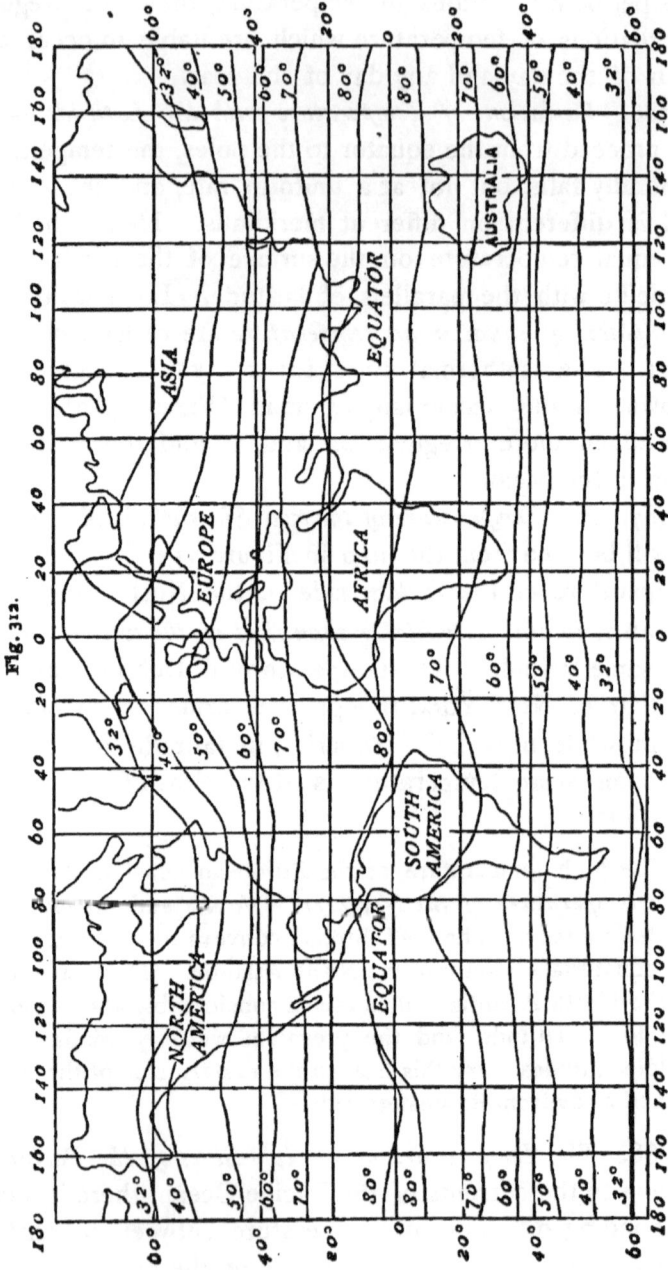

Fig. 312.

higher than that of the west. This causes a marked difference of temperature between the eastern and western coasts of North America at places on the same parallel. The same isothermal line will be found 10 or 15 degrees farther north on the Pacific coast of North America than on the Atlantic coast.

389. *The Temperature of the Northern and Southern Hemispheres.* — The mean temperature of the *northern* hemisphere is *nearly three degrees higher than that of the southern hemisphere.*

The unequal temperature of the two hemispheres is probably due to *the unequal distribution of land and water.* The northern hemisphere contains more land and less water than the southern. In the southern hemisphere the sun's rays fall chiefly upon water, and a large amount of heat is consumed in evaporation. In the condensation of vapor the heat is again liberated. Observations show that there is more condensation in the northern hemisphere than in the southern. Thus the soutnern hemisphere is cooled more by evaporation and warmed less by condensation than the northern hemisphere.

390. *Mean and Extreme Temperatures of a Place.* — Two places having the same mean temperature may differ greatly in their extreme temperatures. New York and Liverpool have the same mean temperature, but the difference between the mean temperature of the three summer months and that of the three winter months is twice as great in New York as in Liverpool.

In some localities the mean temperature of the hottest month of the year is less than 5° above that of the coldest, while in other localities it is 70° or 80° above.

391. *Marine and Continental Climates.* — The temperature of water *changes less than that of land.*

The specific heat (197) of water being much higher than that of land, a much greater amount of heat is consumed in raising the temperature of an equal mass of water the same number of degrees, and a much greater amount of heat is liberated in the cooling of an equal mass of water. Hence when land and water are receiving or losing heat at the same rate, the temperature of the former will rise higher or fall lower than that of the latter in the same time. The high latent heat of watery vapor (202) tends to keep the temperature of water uniform, a large amount of heat being rendered latent by evaporation when the temperature is rising, and an equally large amount being liberated by condensation when the temperature is falling. Again, the sun's rays penetrate water to a greater depth than land, and at the same time the currents in the ocean tend to equalize the temperature of the water at different depths. Hence, while land becomes heated only at the surface, water becomes heated to a considerable depth below the surface. The greater depth of water heated and cooled as the temperature rises and falls would cause the temperature to change less at the surface of water than of land.

When the temperature of a place is controlled mainly by the *ocean*, the temperature is *equable*, and the climate is called *marine;* when, on the contrary, it is controlled mainly by the *continent*, the temperature is *extreme*, and the climate is called *continental*. On the eastern coast of the United States, where the prevalent winds are from the land, there is a great annual range of temperature and a continental climate ; while in the western part of Europe, where the prevalent winds are from the ocean, the temperature is more uniform and the climate marine.

392. *Change of Temperature with the Elevation.* — As we *ascend* in the atmosphere from the earth, *the temperature falls*. The rate of decrease varies with the latitude of the place, with the time of the year, and with the hour of the day. It is more rapid in warm countries than in cold, and in the hot months than in the cold. It is most rapid about 5 P. M., and least rapid about sunrise. The change is also most rapid near the earth, and decreases as we ascend.

There are two main reasons why the temperature of the atmosphere falls as we ascend : (1) The air of the earth's surface becomes heated and expanded, and tends to rise because of its diminished specific gravity. As the air ascends it meets

with less pressure and therefore expands ; this expansion consumes heat (204), and causes the temperature to fall. (2) The
moisture in the air becomes less and less as we ascend, and
hence there is less absorption of the solar rays, and it is only
the rays which are absorbed that tend to raise the temperature ;
there also is less hindrance to the escape into space of the heat
radiated from the atmosphere.

393. *The Line of Perpetual Snow.* — Since the temperature of the atmosphere falls as we ascend, the tops of
high mountains, even 'within the tropics, are covered with
perpetual snow. The snow-line depends more upon the
temperature of the hottest month than upon the mean temperature of the year. It is not therefore the line whose
mean temperature is 32°. It depends also to a considerable extent upon the annual snow-fall.

Under the equator the height of the snow-line varies from
15,000 to 16,000 feet, where the mean annual temperature is
35°. On the Alps the average height of the snow-line is 8800
feet, where the mean annual temperature is 25° ; while on the
coast of Norway its height is only 2400 feet, where the mean
annual temperature is 21°.

394. *The Atmosphere a Regulator of Temperature.* —
During the day the atmosphere absorbs a portion of the
sun's rays, so that they are less excessive on reaching the
earth. A considerable portion of the heat thus absorbed
during the day is rendered latent by expansion. At night
the air intercepts a part of the rays emitted by the earth,
and so keeps the heat from escaping into space. At the
same time, as the air is cooled, it contracts, and so liberates
the heat that was rendered latent by expansion during the
day. Were it not for the atmosphere the days would be
very much hotter and the nights very much colder than
they are now. It is chiefly by means of the *watery vapor*
present in the atmosphere that it acts thus as a regulator
of temperature (381).

III.

HUMIDITY OF THE ATMOSPHERE.

395. *The Hygrometer.* — An instrument capable of *measuring the moisture of the air* is called a *hygrometer*. A *hygroscope* is an instrument which merely shows that there are *changes of moisture,* without being capable of measuring the *amount* of moisture present.

Mason's hygrometer (Figure 313) consists of two thermometers. The bulb of one of these is kept moist by being covered with muslin or silk, the fibres of which dip into a reservoir of water. The water is drawn up to the bulb by capillary action, and the evaporation from its surface lowers its temperature. Hence the wet-bulb thermometer will always show the lower temperature. The greater the difference of reading between the thermometers, the faster the evaporation from the wet bulb and the drier the air.

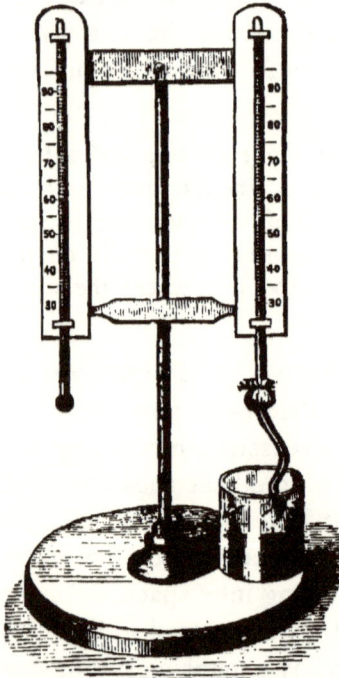

Fig. 313.

396. *The Humidity of the Air.* — The amount of moisture which a cubic foot of air can hold *increases with the temperature.* When the air *contains all the moisture it can hold* at that temperature, it is said to be *saturated* with moisture. By the *humidity* of the air we do not mean the absolute amount of moisture in it, but its *degree of saturation.* If the air is half saturated, its humidity is 50; if three-quarters saturated, 75 ; etc.

397. *The Dew-Point.* — The *dew-point* is the temperature at which the air would become *saturated* with the moisture in it, and its moisture *begin to be deposited as dew.* It is not a fixed temperature, like those of the freezing and boiling points, but *varies with the temperature and humidity* of the air. The greater the humidity of the air, the less the temperature would have to fall to reach the dew-point.

398. *Diurnal Variation in the Vapor in the Atmosphere.* — The amount of vapor in the atmosphere is subject to great fluctuations, some of which are irregular and others periodic. As a rule, the amount is *least about an hour before sunrise,* and *greatest just before sunset,* the mean diurnal variation amounting to about ⅛ of the average amount of vapor present.

The curve in Figure 314 shows the diurnal variation at Philadelphia, the figures at the left indicating the pressure of the vapor in inches of mercury at the hours given at the bottom. This variation is due to the diurnal change in tempera-

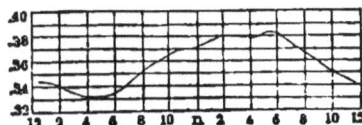

Fig. 314.

ture. As the temperature rises during the day, more water is evaporated from the ocean and the moist earth, and the amount of vapor in the air increases. During the night a portion of the vapor is condensed in the form of dew and hoar-frost, and the amount present in the air decreases.

399. *Annual Variation in the Amount of Vapor in the Atmosphere.* — In the northern hemisphere the mean amount of vapor in the atmosphere is *greatest in July,* when the mean temperature is highest, and *least in January,* when the mean temperature is lowest. This is due to the more rapid evaporation in summer than in winter.

400. *Variation in the Amount of Vapor with the Elevation.* — The humidity of the atmosphere as a rule *decreases as we rise* above the earth, though there is a slight increase of humidity for the first 3000 feet. At the highest elevations at which

observations have been taken the air has never been found entirely free from moisture.

401. *Diurnal Variation of the Pressure of the Gaseous Atmosphere.* — The earth is really enveloped in *two atmospheres*, one of *vapor* and one of *permanent gases.* These two atmospheres are mixed together, and by their combined pressure cause the rise of the barometer. Other things being equal, *the greater the amount of vapor* present in the atmosphere *the higher the barometer,* and vice versa. Fluctuations in the height of the barometer are caused by changes in the temperature of the air and the amount of vapor present in the atmosphere. A diminution of vapor and an increase in temperature both tend to cause the barometer to fall.

If we subtract the pressure of the vapor in the atmosphere from that of the whole atmosphere, the remainder will be the

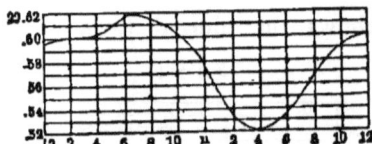

Fig. 315.

pressure of the gaseous atmosphere. At Philadelphia this pressure is greatest about an hour after sunrise and least about 4 P. M., as is shown by the curve of Figure 315.

402. *Annual Variation of the Pressure of the Gaseous Atmosphere.* — In the northern hemisphere the pressure of the gaseous atmosphere is *greatest in January,* when the temperature is lowest, and *least in July,* when the temperature is highest. The difference between the summer and winter pressures of the gaseous atmosphere is very unequal in different countries. In the eastern part of the United States this difference amounts to about half an inch, while in Central Asia it amounts to above an inch, and at the equator is scarcely appreciable.

IV.

MOVEMENTS OF THE ATMOSPHERE.

403. *Winds.* — Wind is *air in motion.* Although the winds are proverbially variable and fickle, they are gov-

erned by laws as fixed and definite as those which regulate the temperature and pressure of the atmosphere.

The force of a wind is estimated either by its *velocity in miles per hour* or by its *pressure in pounds per square* foot. The character, velocity, and pressure of various winds are given in the following table, taken from Loomis : —

Character.	Velocity in Miles per Hour.	Force in Pounds per Square Foot.
Just perceptible	2	0.02
Gently pleasant	4	0.08
Pleasant brisk	12½	0.75
Very brisk	25	3.00
High wind	35	6
Very high wind	45	10
Strong gale.	60	18
Violent gale	70	24
Hurricane	80	31
Most violent hurricane	100	49

From a long series of observations at Philadelphia, it appears that the mean velocity of the wind is 11 miles an hour. The mean velocity varies somewhat during the day and during the year. It is least about sunrise and greatest about 2 P.M. It is

Fig. 316.

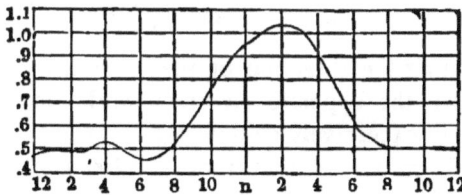

nearly uniform during the night. The curve in Figure 316 shows this diurnal variation in the force of the wind, the figures in the vertical line indicating the pressure in pounds per square foot.

404. *The Cause of Winds*. — Movements of the atmosphere are produced either by the *unequal pressure* of the atmosphere at different points, or by the *unequal specific gravity* of different portions of the atmosphere.

Surface currents will always set in *from* a region of *high* pressure *towards* a region of *low* pressure.

Unequal specific gravity of the air may be due to *inequalities of temperature or of humidity*.

Suppose the surface of the earth in the neighborhood of *C*

Fig. 317.

(Figure 317) to become excessively heated. The air above *C* will by expansion become lighter than the surrounding air. This lighter air will accordingly rise, and its place will be supplied by an inflow along the surface from every side. At the same time the heated column, rising above the surrounding atmosphere, gives rise to an outflow at the top. At a certain distance from the heated column there will be descending currents to supply the place of the air which is flowing in towards the heated region at the surface of the earth.

Fig. 318.

The system of currents that would be developed on every side of an excessively heated region is shown in Figure 318, the arrows indicating the direction of the currents. A system of currents in just the opposite direction would be developed on every side of an excessively cold region.

The specific gravity of the vapor of water is only about two thirds that of dry air. As it takes time for the vapor to diffuse itself into the atmosphere, an excess of aqueous vapor tends to produce a region of low specific gravity, and so to develop a system of currents similar to those developed by a region of high temperature.

405. *The Direction of the Winds modified by the Rotation of the Earth.* — The earth's rotation from west to east in 24 hours is on an axis perpendicular to its equator. Every point on the earth's surface is carried around in the same time, but points near the equator describe longer paths, and hence must move with greater velocity than those near the poles. The velocity of rotation at the surface is greatest at the equator and decreases towards the poles. At the equator it is 1036 miles per hour ; 15° from the equator it is 1000 miles per hour ; 30° from the equator, 897 miles ; 45° from the equator, 732 miles ; 60° from the equator, 518 miles ; 75° from the equator, 268 miles.

If a mass of quiescent air from parallel 30° were suddenly transported to parallel 15°, it would have an easterly motion of 103 miles an hour *less* than that of the parallel arrived at. It would therefore seem to be moving over the surface of the earth westward at the rate of 103 miles an hour. Of course it would really be the surface of the earth which would be moving under it eastward at that rate. The effect upon bodies on the surface of the earth would be the same as if the earth was stationary, and the wind blowing over it to the west at the above rate.

If, on the other hand, a mass of quiescent air were suddenly transported from parallel 15° to parallel 30°, it would have an easterly motion of 103 miles an hour *greater* than the parallel arrived at.

In general, any wind blowing *towards* the equator is deflected towards the *west* by the rotation of the earth, so as to make it an *easterly* wind ; and any wind blowing *from* the equator is deflected towards the *east* by the rotation of the earth, so as to make it a *westerly* wind.

406. *Systems of Winds.* — There are three great systems of winds upon the globe, namely, the *trade-winds*, the *middle-latitude winds*, and the *polar winds*.

407. *Trade-Winds.* — There is a belt of excessively heated air surrounding the earth within the tropics. This heated air develops a system of currents on each side of it, similar to those described in section 404. Surface currents set in *towards* the equator *from* the north and the south, and upper currents *from* the equator *towards* the north and the south. The rotation of the earth deflects the surface currents towards the west, so as to make them *easterly* winds ; and the upper currents towards the east, so as to make them *westerly* winds. The surface wind north of the equator becomes a *northeast* wind, and that south of the equator a *southeast* wind. These winds are called *trade-winds*, from the service they render to commerce.

408. *Cause of the High Barometer near the Parallel of* 32°. — As the upper equatorial currents move towards the poles they tend to increase the pressure of the atmosphere towards the north and the south ; for since the meridians converge as we proceed from the equator towards the poles, the air as it moves towards the poles must increase in depth, and so produce a greater pressure at the surface. This increased pressure of the air in middle latitudes arrests the further progress of the polar current, and a calm ensues. The upper air descends to the earth's surface, and joins the surface current towards the equator, where it again ascends, and thus maintains a perpetual circulation.

409. *The Middle-Latitude Winds.* — The high pressure near the parallel of 32° gives rise to *surface currents from the equator towards the poles*, and to *upper currents from the poles towards the equator*. The *surface* currents are *deflected by the rotation of the earth* towards the east, so as to make them *westerly* winds ; and the *upper* currents *towards the west*, so as to make them *easterly* winds. These surface currents are the prevailing winds of the middle latitudes. In the *northern* hemisphere they blow from a point *a little south of west*, and in the *southern* hemisphere from a point *a little north of west*. Throughout the middle lati-

tudes of the United States the average direction of the wind is 10° south of west; and the easterly winds are to the westerly as 2 to 5. In corresponding latitudes in the southern hemisphere the prevalent direction of the surface winds is 17° north of west; and the easterly winds are to the westerly as 1 to 5.

410. *The Polar Winds.* — The extreme cold of the polar regions produces the opposite effect to that of the extreme heat of the tropics. It produces great density of air, and develops *surface currents from the poles towards the equator,* and *upper currents in the opposite direction.* These currents are *deflected by the rotation of the earth,* as in all other cases. The polar and middle-latitude winds encounter each other near the parallel of 60°.

Fig. 319.

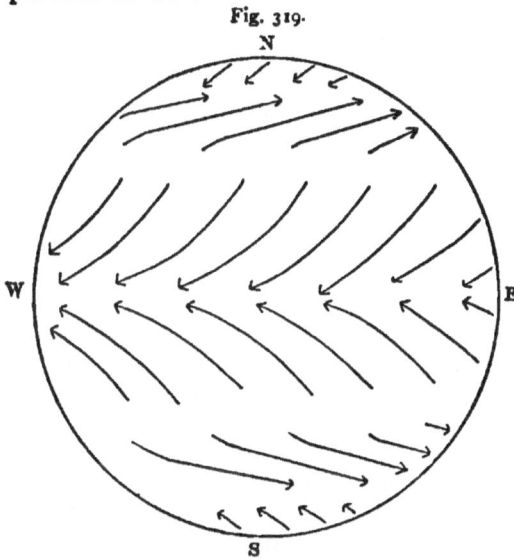

The three systems of surface winds are shown in Figure 319, the arrows indicating the direction of the wind in each belt. Figure 320 shows the complete circulation of the atmosphere.

411. *Monsoons.* — During the summer months the surface of the land becomes heated to a higher temperature than

that of the surrounding water, while during winter it becomes
cooled to a lower temperature (391). Hence during the
summer months there is a ten-
dency to develop *surface winds
from the oceans to the continents,*
and *in the opposite direction* during
the *winter* months. This tendency may either give rise to winds
in the direction in which it acts,
or merely modify the direction and
force of the prevailing winds.

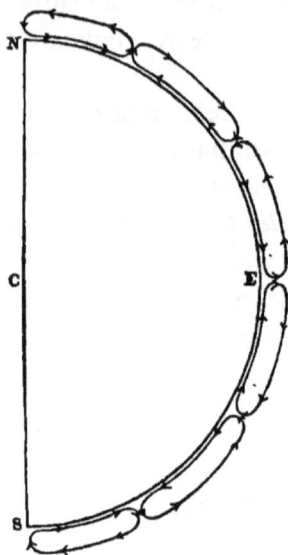

Fig. 320.

In the former case we have
what are called *monsoons*, that is,
winds which *blow during the summer months from the water to the
land, and during the winter months
from the land to the water.* The
most marked monsoons on the
globe are those on the south coast
of Asia, in the region of the northeast trades. The tendency of the unequal heating of the continent of Asia and
of the Indian Ocean during the winter months is to produce
a wind in the direction of the trade-wind, and in the
summer months in the opposite direction. The winter
monsoon adds to the force of the trade-wind, while the
summer monsoon overbalances the trade, and produces
a wind in the opposite direction.

412. *Land and Sea Breezes.* — During the day the surface of
the land becomes hotter than that of the neighboring water, and
at night cooler. There is, therefore, a general tendency for the
wind to blow from the water to the land during the day, and
from the land to the water at night. When this tendency is
strong enough to produce a wind in the direction in which it
acts, we have what are called *land* and *sea breezes*, or winds
blowing *from the sea during the heat of the day*, and *from the
land during the cool of the night.* These winds are strongest
on islands in tropical regions.

V.

CONDENSATION IN THE ATMOSPHERE.

A. Dew and Hoar-Frost.

413. *Origin of Dew.* — All bodies on the surface of the earth are radiating heat to the sky, and when they thus part with heat faster than they receive it, their temperature falls below that of the surrounding air. When the sun is above the horizon, they generally receive heat faster than they part with it by radiation, but *at night* they usually *radiate heat faster than they receive it.*

When the blades of grass, leaves of plants, and other objects on the surface of the earth become cooled by radiation below the dew-point of the atmosphere, *they condense upon themselves a portion of the atmospheric moisture* in the form of *dew.* The greatest amount of dew is deposited upon the substances whose temperature becomes the lowest. Dew does not fall from the sky like rain, but collects upon those bodies which are cool enough to condense the vapor in the air in contact with them. A pitcher of ice-water, on a warm summer's day, becomes quickly covered with a film of dew, the cold surface of the pitcher condensing the vapor from the layer of air in contact with it.

414. *Circumstances favorable to the Formation of Dew.* — Anything which *favors the loss of heat by radiation* is favorable to the formation of dew.

A *cloudless night* and an unobstructed *exposure to the sky* are especially favorable to the formation of dew, because they allow the heat radiated by bodies to escape freely into space. A cloudy night or any artificial covering, however slight, prevents the formation of dew, for the clouds or coverings reflect back the heat radiated from the earth, and so keep bodies on its surface from cooling below the dew-point.

A *slight breeze* favors the formation of dew by renewing the air in contact with the surface as fast as it deposits its excess of vapor. A stiff breeze, however, prevents the formation of dew by allowing no layer of air to remain long enough in contact with the surface of a body to become sufficiently cooled to deposit its moisture. There is little dew on windy nights.

A *moist atmosphere* favors the formation of dew, because the more moisture in the air the less the reduction of temperature at which the deposition of dew will begin. *Good radiators and bad conductors receive the greatest amount of dew.* The temperature of their surfaces falls rapidly at night, because they lose heat rapidly by radiation and receive it slowly by conduction from their interior or from the earth with which the bodies are in contact. Wool, being a good radiator and a poor conductor, collects a large amount of dew at night, while a plate of polished metal will receive scarcely any at all.

415. *Formation of Hoar-Frost.* — When the temperature of the surface is below the freezing-point, the moisture of the atmosphere is deposited upon it *in the solid state*, as *frost.* Hoar-frost is not frozen dew, but *frozen vapor*, that is, vapor deposited in the solid form without passing through the liquid state.

Since the leaves of plants sometimes become cooled by radiation several degrees below the air a few feet from them, there may be a frost when the thermometer indicates a temperature several degrees above the freezing-point. There is not, however, likely to be a frost unless the temperature of the dew-point is below 32°. The temperature of the surface will not fall much below the dew-point, because of the heat which is liberated on the deposition of the dew.

416. *Frost in Valleys.* — There is often sufficient frost in valleys and up to a certain height on the hillsides to kill plants, while higher up there is no frost at all. As the air on the hillsides is cooled by contact with the cold surface, it gradually

settles into the valley, becoming cooler and cooler by contact
with the surface as it descends, and raising the warmer air
bodily out of the bottom of the valley, just as a heavy liquid
will raise a lighter one by flowing under it.

B. Fog and Mist.

417. *Origin of Fog.* — The watery vapor of the atmosphere is transparent, but when from any cause a portion
of the atmosphere becomes cooled below the dew-point, a
part of the vapor becomes condensed into *minute drops of
water which float in the atmosphere.* The partially condensed vapor becomes visible as a *mist* or *cloud.* When
the condensation takes place *near the surface of the earth* it
gives rise to a *fog* or *mist.*

When steam rises from a vessel of warm water and mixes
with the colder air above, a portion of the vapor is condensed
into a mist which is often improperly called steam. Steam
proper is a *gaseous* body, while mist is a *liquid* body.

418. *Fogs over Rivers.* — At certain seasons of the year,
and especially during the latter part of the summer, fogs
form over rivers and lakes almost every clear and still
night. During the night the air over the land becomes
cooler than the water of the lake or river, and as the vapor
rises from the water it is *partially condensed by contact with
the cooler air from the land,* and gives rise to a fog which
floats upon the surface of the water. Fogs are often
formed in a similar manner over harbors and bays, and
these fogs are frequently drifted inland by gentle currents
of air.

419. *Fogs on the Breaking up of Frost.* — Extensive fogs
often occur in midwinter after a thaw or a warm rain. In
this case *warm and moist currents of air become chilled in
passing over the cold surface of the frozen ground,* and a part
of the moisture is condensed as a fog. For a similar
reason icebergs are liable to be enveloped in mist, the ice

cooling the surrounding air sufficiently to condense a part of its moisture.

420. *Fogs on the Banks of Newfoundland.* — Fogs prevail along the northern side of the Gulf Stream, the warm and moist air over the Gulf Stream being chilled by contact with the colder air from the water on the north. These fogs are especially prevalent over the Banks of Newfoundland. They occur every month of the year, but are especially frequent in the summer, when the Banks are enveloped in fog nearly half of the time. The shallow Banks compel the cold arctic current at the bottom of the ocean to come to the surface, and the cold water thus brought to the surface chills the air laden with moisture from the Gulf Stream.

421. *Mist on the Tops of Mountains.* — The tops of mountains are liable to be enveloped in mist. The mountains compel the warm currents of air to rise to pass over them. As these currents rise they become chilled partially by expansion (204), and partially by contact with the cold surface of the mountains. When the air is chilled below its dew-point, a mist is formed, which is again dissipated as the air passes down into warmer regions on the other side of the mountains.

422. *How Fog is sustained in the Air.* — The particles of fog are sustained in the air *in the same manner as a cloud of dust.* The dust remains for a long time suspended in the air, although each particle may consist of matter two thousand times as dense as the air in which it floats. When the air is perfectly tranquil, these particles do indeed fall, but their descent is so slow that their motion is perceptible only after a considerable interval of time.

423. *The Indian Summer.* — At certain seasons of the year there occurs what is called a *dry fog.* In the United States this frequently occurs in November, or the latter part of October, and this period is commonly known as the *Indian Summer.* It is characterized by a hazy atmosphere, a redness of the sky, absence of rain, and a mild temperature. This appears to result from a dry and stagnant state of the atmosphere, during which the air becomes filled with dust and smoke from numer-

ous fires. A heavy rain washes out these impurities and clears the sky. Long periods of drought in summer are characterized by a like condition of the atmosphere.

C. Clouds and Rain.

424. *Nature and Formation of Clouds.* — A cloud differs from a fog simply in its elevation above the earth. A *fog* might be defined as *a cloud resting on the earth;* and a *cloud,* as *a fog floating in the air.*

Clouds are formed whenever *a mass of air away from the earth's surface is cooled below its dew-point.* This cooling may be effected in various ways. A cold wind may penetrate a mass of warm air and cool it below its dew-point, or a warm moist wind may be thus cooled by penetrating a mass of cold air. Ascending currents of air are always cooled by expansion (204), and are very likely to give rise to clouds.

425. *Clouds on the Summits of Mountains.* — The summits of high mountains are usually enveloped in clouds even when the rest of the sky is clear. An interposed mountain forces a horizontal wind up to an unusual height where the temperature is low. When the temperature of the ascending current reaches its dew-point, a portion of its moisture is condensed as a cloud. Let *A B C* (Figure 321) be a mountain interposed in the path of a horizontal current. The current will be forced upward, and made to glide along the side of the mountain. Let *D E* represent the elevation at which the temperature of the ascending current will just reach its dew-point. As soon as the current passes above this line its vapor will be partially condensed so as to form a cloud, which will envelop the summit of the mountain. As soon as the current passes below the line *D E* on the other side of the mountain, its temperature again

Fig. 321.

rises above its dew-point, and the cloud is redissolved. The cloud is drifted by the wind, but is not blown away from the mountain because, as fast as it moves forward, a new cloud is formed behind it. Although the cloud on the mountain appears stationary, *the particles which compose it are continually changing.*

Even in tolerably level countries, the sky does not become overcast solely by clouds drifted by the wind from places beyond the horizon. The clouds are *very often formed in sight of the observer.* So too the sky often clears, not because the clouds are drifted off by the wind, but because they are *dissipated by the increasing heat* of the air.

Fig. 322.

426. *The Classification of Clouds.* — The four chief varieties of clouds are the *cirrus,* the *cumulus,* the *stratus,* and the *nimbus.*

The *cirrus* cloud consists of *long, slender filaments, either parallel or diverging,* and often resembling a lock of cotton whose fibres are electrified so as to repel one another. These clouds have the least density, the greatest elevation, and the greatest variety of form. They are generally the first to appear after a period of perfectly clear weather. They are believed to be composed of spiculæ of ice or flakes of snow floating at a great height in the air. At the height at which they prevail the temperature of the air is below 32° even in midsummer (Figure 322).

The *cumulus* cloud usually consists of *a rounded mass, rising from a horizontal base.* It is much denser than the cirrus, and is formed in the lower regions of the atmosphere. In fair weather it often forms a few hours after sunrise, increases until the hottest part of the day, and disappears about sunset. We often see near the horizon large masses of this cloud which resemble mountains covered with snow.

The rounded top of the cumulus is due to the mode of its formation. When the surface of the earth is heated by the rays of the sun, currents of warm air ascend, and at a certain height a portion of their vapor is condensed into cloud ; and since the upward motion is greatest under the centre of the cloud, the vapor is there carried up to the greatest height (Figure 323).

Fig. 323

The *stratus* cloud is *a widely extended horizontal sheet,* often covering the sky with a nearly uniform veil. It is the lowest of the clouds, and sometimes descends to the surface of the earth (Figure 324).

The *nimbus* is the well-known *rain-cloud,* consisting of a combination of cirrus, cumulus, and stratus clouds (Figure 325).

427. *The Height and Thickness of Clouds.* — The height

of clouds is very variable. The stratus cloud sometimes descends to the surface of the earth. In pleasant weather the under surface of the cumulus cloud is from 3000 to 5000 feet high. Cirrus clouds are never seen below the summit of Mont Blanc.

Fig. 324.

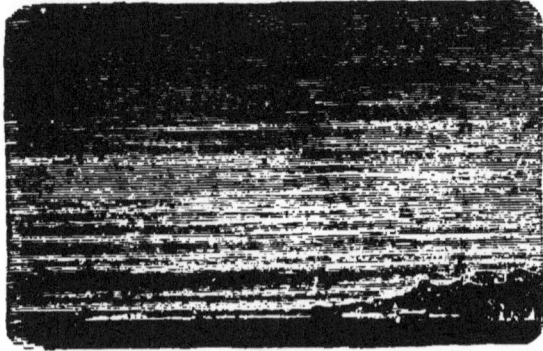

Clouds are not usually more than half a mile thick, though cumulus clouds sometimes attain a thickness of 3 or 4 miles.

Fig. 325.

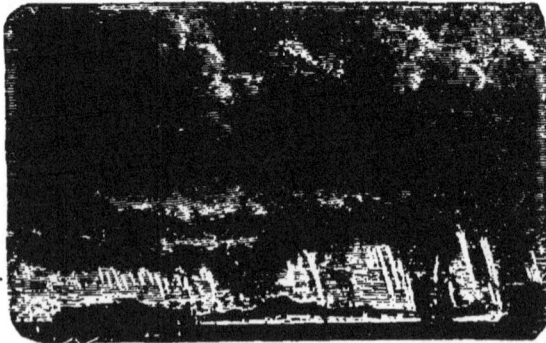

428. *How Clouds are sustained in the Atmosphere.* — Since clouds are composed of particles heavier than air, they must be slowly sinking. They do not ultimately fall to the earth in pleasant weather, because, as they sink,

they encounter warmer layers of air which are not saturated with vapor. The cloud is therefore *dissipated at the bottom as fast as it falls,* while it is at the same time *renewed at the top by the condensation of vapor* carried up by ascending currents.

429. *Origin of Rain.* — Rain has the same origin as clouds. *When the condensation takes place slowly, clouds only are formed;* but *when it takes place with sufficient rapidity, rain is also formed.* To produce an abundant rain, the air must be suddenly cooled below the dew-point. The most effective way to accomplish this is to force the air up a mile or two above the surface of the earth. Were a mass of air raised two miles from the surface of the earth, its temperature would fall about 35°, partly from the chilling effects of expansion and partly from the coldness of the space into which the air would be transported. Were the air of the surface of the earth forced up to this height, most of its vapor would be condensed. The air may be forced upward by the interposition of a mountain range in the path of a current of air, or by the meeting of two opposing currents. Hence *mountain ranges are very efficient condensers of the atmospheric vapor.* The heaviest rainfall on the globe occurs where the prevailing wind is from the ·ocean, and is obliged to pass over a high mountain range on its way to the interior of the continent. On the shel tered side of such ranges there are often desert regions.

430. *The Amount of Rain in Different Latitudes.* — The average rainfall is *greatest at the equator,* and *decreases towards the poles.* The annual rainfall at the equator is 104 inches; in latitude 20°, 70 inches; in latitude 30°, 40 inches; and in latitude 60°, 20 inches.

431. *Origin of Snow.* — Snow bears the same relation to rain that hoar-frost does to dew. When the vapor of the atmosphere is precipitated at a very low temperature, it at once *assumes the solid state,* usually in the form of

minute crystals, which attach themselves to each other and form *snow-flakes*, which fall slowly to the earth. Snowflakes present a great variety of forms, some of which are shown in Figure 326.

Fig. 326.

432. *Hail*. — Large hail seldom if ever falls except during thunder-storms. It very rarely follows rain which

Fig. 327.

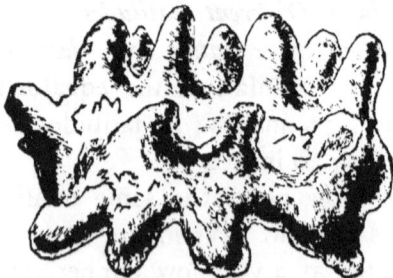

has continued for some time. The hail covers a much smaller area than the rain-storm, and usually continues at the same place for only five or ten minutes. Hailstones are of all sizes, from that of small shot to that of a turkey's egg, and of every variety of shape.

One of very irregular form is shown in Figure 327. The centre of large hailstones usually consists of hardened snow, and this is surrounded by a layer of transparent ice. Sometimes we find several alternate layers of opaque snow and transparent ice. Figure 328 shows

Fig. 328.

the section of a hailstone whose external form is given in Figure 329.

Fig. 329.

432a. *Origin of Hail.* — The formation of hail is invariably attended by two distinct currents of air, one of which displaces the other with great violence. The current of air which precedes the approach of a hail-storm is extremely hot, and highly charged with moisture ; and that which succeeds the fall of hail has an icy chillness. The warm and humid air is displaced by the cold current, and is thus forced up to a great height, by which means its vapor is suddenly condensed. Upon the front of the hail-cloud this condensed vapor exists in the form of water, whose temperature is near 32°. In the interior of the cloud the vapor is precipitated in the form of snow, whose temperature may be as low as 20°.

Observations on mountains have shown that, on the front of the hail-cloud there is a violent whirling motion about a horizontal axis. This causes the snow to collect in small balls, each of which forms the nucleus of a hailstone. The snow-ball is forced into the warm current, where it receives a layer of water, which is congealed by the nucleus, thus rendering the snowy centre more compact, and adding a shell of transparent ice. By the whirling motion, the hailstone is then hurled into the snow-cloud, where it receives a layer of snow, and again becomes thoroughly chilled. Thence it escapes again into the water-cloud, and is covered with a layer of water, which is frozen by the cold of the nucleus. Thus it is plunged alternately into the snow-cloud and the water-cloud, while each alternation furnishes a layer of spongy ice and a layer of transparent ice. Hence the stone grows rapidly, and in a few minutes becomes a ball three or four inches in diameter.

The hailstones are sustained in the air by the violent upward motion caused by the cold current displacing the warm one. A sphere of ice two inches in diameter, falling through a tranquil atmosphere, soon acquires a velocity of 90 feet per second. A hailstone of irregular shape would experience more resistance than a sphere, and would acquire less velocity, but it would still fall from a height of 18,000 feet in about three minutes, which time is too small to allow the formation of masses of ice weighing a pound. An upward current of air, rising with a velocity of 90 feet per second, would sustain a sphere of ice two inches in diameter, and would greatly reduce the velocity of stones of larger size.

D. Storms.

433. *Origin of Storms.* — Any *violent and extensive commotion of the atmosphere* is called a *storm.* Such commotions are usually attended by a fall of rain, snow, or hail, but the storm often extends beyond the area of snow or rain, and even beyond the area of clouds.

Storms are caused by a *strong and extensive upward motion of the air.* Since the air is heated by contact with the earth, and by absorption of solar and terrestrial radiations by the watery vapor it holds, (381) it becomes heated chiefly at the bottom, the watery vapor existing chiefly in the lower layers. An excessive heating of a mass of air at the surface of the earth gives rise to a system of currents such as has been already described (404). A vertical section of this system of currents is shown in Figure 330; a horizontal section at the bottom, in Figure 331; and a horizontal section at the top, in Figure 332.

Fig. 330.

As the air in the centre of the area rises, it is cooled by expansion at the rate of about 38° for every two miles of ascent. The height to which the air will have to rise to be cooled to its dew-point depends upon the difference between the dew-point and the temperature of the air. As soon as the cloud begins to form, the latent heat of the vapor is set

Fig. 331.

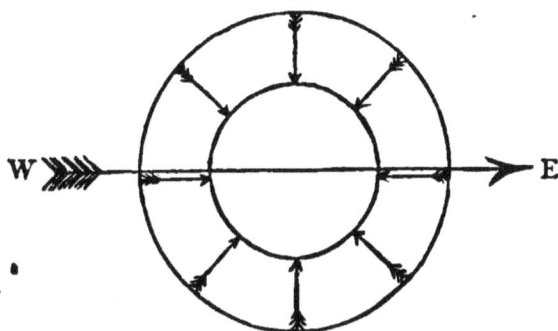

free (139). A rainfall of one inch precipitates over two million cubic feet of water upon one square mile of surface, and liberates as much heat as it would take to evaporate two million cubic feet of water. It takes over 60,000 units of heat (196) to evaporate one cubic foot of water. The heat thus liberated warms the air in the region in which the condensation occurs, and causes the mass of air to rise still higher, so that more of its vapor is condensed and more heat is liberated.

Fig. 332.

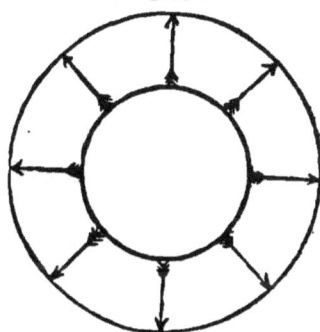

The expansion of the column of air ascending in the centre of a storm, especially after heat begins to be liberated by the condensation, causes the air to spread out in all directions above, making a barometer under the centre of the cloud fall below its mean height, and one beyond the limits of the cloud rise above its mean height. Near the limits of the cloud the air, being heavier, sinks downward, and a portion of it flows along the surface towards the centre of the ascending column, while another portion flows along the surface in the opposite direction, producing a gentle breeze away from the cloud. The air spreads out more rapidly above than it runs in below, and the storm tends to increase in diameter. Storms often extend with great rapidity till they cover an area of more than a thousand miles in diameter.

434. *The Development and Motion of Storms.* — Storms begin gradually, and are usually a day or two in attaining their greatest violence. After a day or two longer their violence again decreases, and at length they disappear or are merged into other storms. A storm occasionally lasts one or two weeks, but usually only a few days. It sometimes remains nearly stationary for a day or two, but it usually moves eastward about 600 miles a day.

The average *direction* of storms across the United States is a little north of east, being almost exactly east during the summer.

The average *velocity* of a storm is twenty-six miles an hour, being twenty-one in the summer and thirty in the winter. They occasionally move at the rate of fifty miles an hour, and sometimes remain almost stationary for a day or two.

The direction in which a storm moves is entirely distinct from that of the wind which accompanies it. While the storm moves steadily eastward, the wind has every possible direction at places within the limits of the storm. At places on the north side of the centre of the storm, the wind usually sets in from the northeast as the storm approaches, and veers round by the north to the northwest as the storm passes over. At places on the south side of the centre, the wind generally sets in from the southeast, and then veers round by the south to the southwest.

Near the centre of a great storm there is usually a lull in the wind, and sometimes a calm. There is seldom any rain, and the clouds often break, and occasionally there is a clear sky for several hours. Soon after the centre of the storm has passed the wind changes to the west, and there is a heavy fall of rain or snow of comparatively short duration.

The winds on the east side of a storm are propagated in a

direction opposite to that in which they blow. That is to say, they are propagated *eastward* while they blow westward. Winds propagated, like these, in the opposite direction to that in which they blow are said to be propagated by *aspiration*. The winds on the west of the storm are propagated in the same direction as that in which they blow. Such winds are said to be propagated by *impulsion*.

435. *Cyclones.* — The inequalities of the earth's surface greatly modify the direction of the wind, so that in great storms the movements of the atmosphere often seem very complex. Over the ocean these disturbing causes do not exist, and in violent storms the movements of the air are much more regular and uniform. The motion of the wind is generally *spirally inward* towards the centre of the storm, and such storms are now commonly designated by the term *cyclone*. These storms prevail in the neighborhood of the West India Islands, where they are known as *hurricanes*. They are also common in the China Sea and in the Indian Ocean.

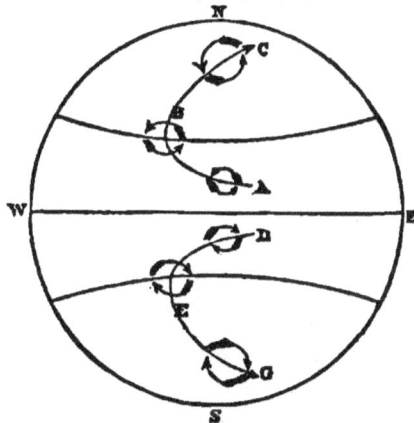

Fig. 333.

Cyclones originate near the equatorial limits of the trade-winds, and move northward and southward in parabolic paths, as shown in Figure 333. The small arrows indicate the direction of the circulation of the wind in the cyclone itself. *Tornadoes* are very violent storms, caused by a sudden and very great fall of pressure.

436. *Predictions founded upon the Established Laws of Storms.* — "The laws of storms are now so well understood that we can predict with some confidence the changes which will succeed at any place during the next few hours, provided we can know the state of the weather throughout the surrounding region to a great distance. This is what has been attempted

since 1871 by the United States Signal Service, and the general accuracy of these predictions has excited considerable surprise. Such predictions would be still more reliable if we could have information respecting the various meteorological elements from a larger portion of the earth's surface. The centre of a large portion of our storms follows nearly the northern boundary of the United States, so that our observations inform us respecting only one half, or perhaps less than one half, of the storm-area. Moreover, storms are often affected by changes which are going on in very distant quarters. An area of unusually high barometer may affect the course of a storm whose centre is distant two or three thousand miles ; and an unusual fall of rain in the equatorial regions may cause an unusual overflow of air to the middle latitudes, resulting in serious disturbances of atmospheric pressure. When the laws of storms have been more precisely defined, and telegraphic reports can be received from a more extended area, we shall doubtless be able to predict coming storms with greater precision."

VI.

ELECTRICAL PHENOMENA OF THE ATMOS-PHERE.

A. Atmospheric Electricity.

437. *Electrical Condition of the Atmosphere.* — The atmosphere is *almost always charged with electricity*, and usually with *positive* electricity. There are, however, great variations in the intensity of the charge, and clouds are frequently charged with negative electricity.

The intensity of atmospheric electricity varies regularly with the hour of the day and with the season of the year. During the day it is least intense at 4 A. M. and at 4 P. M., and most intense at 10 A. M. and at 10 P. M. It is least intense during the summer months and most intense during the winter months. The intensity also increases with the altitude above the surface of the earth.

When the sky is covered with clouds there are frequent changes in kind as well as in intensity of electricity, the atmosphere being sometimes positive and sometimes negative. It is seldom negative, however, except when rain is falling. When snow is falling, the lower layer of air becomes highly charged with electricity. During a thunder-shower the electricity of the air frequently changes in two or three minutes from positive to negative, and back to positive again, and sometimes half a dozen of these changes occur during a single shower.

B. LIGHTNING.

438. *Lightning.* — Two clouds having opposite electricities attract each other, and when they come sufficiently near, *the two electricities rush towards each other with great violence.* This phenomenon is called *lightning*, and is accompanied by an explosive noise called *thunder.*

Since clouds are very imperfect conductors, when the electricity of one part of a cloud is discharged, the electricity of a distant part is but slightly changed. Thus, a single discharge does not establish a complete electrical equilibrium ; but there is a change in the distribution of the electricities upon the surrounding clouds, and there must be a succession of discharges before the electricity is entirely neutralized. Hence results a succession of flashes of lightning and peals of thunder.

A cloud charged with electricity exerts an *inductive* influence upon the earth's surface immediately beneath it, decomposing its natural electricities, repelling electricity of the same kind, and attracting the opposite kind. Accordingly there will sometimes be a *discharge of electricity from the cloud to the earth.* This charge is usually received by the most elevated objects, such as mountains, hills, trees, spires, high buildings, etc. Trees are particularly exposed to strokes of lightning on account of their elevation, as well as of the moisture they contain, which renders them partial conductors of electricity.

439. *Lightning-Rods.* — Buildings may be protected from injury by the use of *lightning-rods.* These are metallic rods running from the top of the building to the ground.

The rods must not be too small, and their parts must be well connected so as to be in good metallic contact. They should run well into the earth at the bottom, and be carefully pointed at the top. There ought to be several sets of points on the top of the building connected by metallic rods with each other and with the rod that runs to the ground ; and if the building is at all large there ought to be several rods running to the ground, all connected together by metallic rods. If the building has a metallic roof, or there are metallic pipes or other masses of metal in the interior, these should all be carefully connected with the rod. The points facilitate the escape of the electricity from the building and the ground around it when these are acted upon inductively by the cloud. Should the electricity be developed by induction more rapidly than it can escape silently from the points, and a spark discharge should take place, the rod serves as the path of least resistance, and the discharge will take this path rather than pass through the building.

440. *Forms of Lightning.* — Lightning exhibits a variety of forms, which have been designated as *zigzag, ball, sheet,* and *heat lightning.*

Zigzag lightning presents a long, irregular, jagged line of light, like the ordinary spark from an electrical machine (318). This zigzag path is sometimes four or five miles, and perhaps even ten miles in length.

Ball lightning appears like a ball of fire, and is usually accompanied by a terrific explosion. It probably results from a charge of electricity unusually intense, which forces a direct instead of a circuitous passage through the air.

Sheet lightning is a diffuse glare of light, sometimes illuminating only the edges of a cloud, and sometimes spreading over its entire surface. This may be sometimes due to distant lightning which illumines a cloud, while the direct flash is hidden by intervening clouds. Sometimes it may result from a movement of electricity in the interior of a cloud which is a very imperfect conductor, producing an illumination analogous to that observed on a plate of moist glass employed in discharging an electrical machine.

During the evenings of summer the horizon is sometimes illumined for hours by flashes of light unattended by thunder. This is called *heat lightning*, and is sometimes due to the reflection from the atmosphere of the lightning of clouds so distant that the thunder cannot be heard. Sometimes, however, this light overspreads the entire heavens, showing that the electricity of the clouds escapes in flashes so feeble that they produce no audible sound. This may occur when the air is very moist, the air being then a tolerable conductor, and offering just sufficient resistance to the passage of the electricity to develop a feeble light.

441. *Thunder.* — The *light* of lightning proceeds from the a:r, *which is heated white-hot along the line of discharge* by the passage of the electricity. The *thunder* seems to be the *noise produced by the sudden expansion and contraction of this heated air.*

Sound travels only 1090 feet a second, while the transmission of light is nearly instantaneous. Hence the sound does not reach the ear until some time after the flash is seen (144). By observing the interval between the flash and the report, the distance of the point of discharge can be ascertained, sound travelling a mile in about five seconds. Thunder is seldom heard more than ten miles away.

The sound is produced instantaneously at every point along the line of the flash, but, since different parts of the flash are usually at unequal distances from the observer, the sound from different points will reach the ear in slow succession, producing a prolonged *peal* of thunder. The prolonged duration of some peals of thunder is in part due to *echoes*, produced by reflection from the sides of mountains or from clouds.

The variable intensity or *rolling* of thunder is due partly to the zigzag course of the discharge, which often brings several points of the flash equally distant from the observer (the sounds from these points reach the ear simultaneously, and so produce a sound of double and triple the intensity) ; and partly to the unequal distance of different parts of the flash, the sound decreasing in intensity as the square of the distance increases (143). The rolling is in part also the effect of echoes.

Thunder often begins with a rattling sound, followed by a loud peal of variable intensity, and ends with a low rattling sound. This may be due to a discharge like that represented in Figure 334. An observer at E would first hear a rattling sound from the branches $A C$, $A C'$, etc., from the first cloud

Fig. 334.

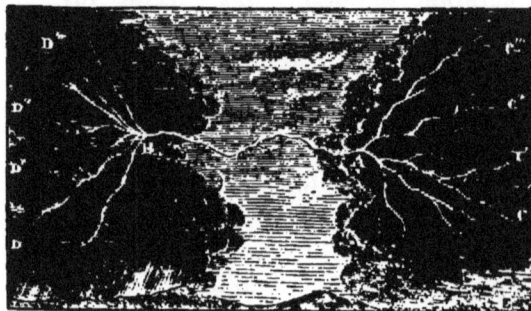

and then a· loud crash of variable intensity from the concentrated discharge between A and B, and finally a rumbling sound from the branches $B D$, $B D'$, etc., of the distant cloud, the noise being feeble on account of the great distance.

C. The Aurora.

442. *The Polar Light.* — The polar light is a luminous appearance frequently seen near the horizon as a *diffused glow*, similar to that of the dawn, whence it has received the name of *aurora*.

443. *Varieties of the Aurora.* — Auroras exhibit a great variety of appearances, but they may be generally referred to one of the following classes : —

1. *A horizontal light like the morning aurora or break of day.*

2. *An arch of light somewhat in the form of a rainbow.* This arch frequently extends entirely across the heavens from east to west, and cuts the magnetic meridian nearly at right angles. It does not long remain stationary, but frequently rises

Fig. 335.

and falls; and several parallel arches are often seen at the same time, appearing as broad belts of light, stretching from the eastern to the western horizon (Figure 335).

Fig. 336.

The arches sometimes present the appearance of a brilliant curtain agitated by the wind (Figure 336).

3. *Slender, luminous beams or columns,* well defined, and often of a bright light. These rise to heights from 20° or 30°

up to 90° or more, sometimes, though rarely, passing the zenith.
Their breadth varies from a quarter of a degree up to two or
three degrees. They often last only a few minutes, but sometimes they continue a quarter or half of an hour, or even
a whole hour. Sometimes they remain at rest, and sometimes they have a quick lateral motion. Their light is commonly of a pale yellow, sometimes reddish, occasionally crimson,
or even blood-red. Sometimes the luminous beams are inter-
spersed with dark rays resembling dense smoke. Sometimes
the tops of the beams are pointed and have a waving motion.

4. Luminous beams sometimes shoot up from nearly every
part of the horizon, and converge to a point a little south of the
zenith, forming *a quivering canopy of flame*, which is called the
corona. The sky now resembles a fiery dome, and the crown
appears to rest upon variegated fiery pillars frequently traversed
by waves or flashes of light. This may be called a *complete
aurora*, and comprehends most of the peculiarities of the other
varieties.

5. *Waves or flashes of light.* The luminous beams sometimes appear to shake with a tremulous motion ; flashes like
waves of light roll up towards the zenith, and sometimes travel
along the line of an auroral arch. Sometimes the beams have a
slow lateral motion from east to west, and sometimes from west
to east. These sudden flashes form an important feature of
nearly every splendid aurora.

VII.

OPTICAL PHENOMENA OF THE ATMOSPHERE.

A. Refraction.

444. *Astronomical Refraction.* — When a ray of light
from a star or other heavenly body enters the atmosphere
obliquely, it will be bent downward, or towards a vertical
line drawn from the point of contact of the ray with the
atmosphere to the surface of the earth ; and as the air
grows denser as we approach the earth, the ray will be
bent more and more as it passes through the atmosphere

from layer to layer (239). As we always see the body which emits the ray in the direction of the ray when it enters the eye, the effect of this refraction will be to *make every heavenly body appear farther above the horizon and nearer the zenith than it really is.* A star in the zenith is not displaced by refraction, because the rays from it enter the air. perpendicularly, and therefore without bending. The farther a star is from the zenith, the more obliquely its rays enter the atmosphere, and the greater the refraction.

445. *Mirage.* — Objects within the atmosphere are sometimes *displaced or made to appear double* by the refraction of the air. This phenomenon is called *mirage.*

Fig. 337.

446. *Mirage upon a Desert.* — Upon a hot desert, on a still day, objects are often seen reflected in a lower stratum of air so as to give *the appearance of water* (Figure 337).

The layers of air near the hot sand become more heated, and consequently rarer, than those higher up. Hence rays coming from any object, as the tree (Figure 338), would, on passing downward, be entering continually rarer and rarer layers of air. They would therefore be bent upward more and more, till they finally meet a layer at an angle exceeding the limiting angle, and become *totally reflected* (240). This total reflection of the rays causes objects to be mirrored in the layers of air as in the surface of water.

Fig. 338.

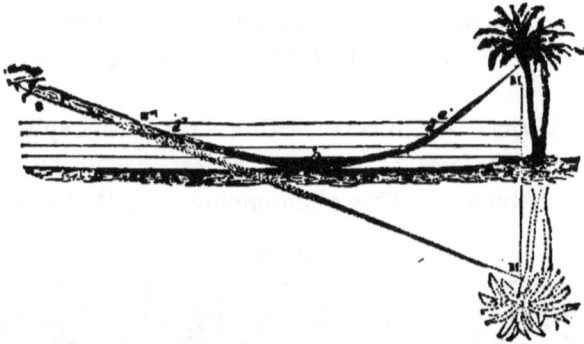

447. *Mirage over Water.* — Objects at a distance over water, partially or entirely below the horizon, often *appear suspended in the air*, sometimes erect, sometimes inverted, and sometimes both erect and inverted, as shown in Figure 339.

Fig. 339.

Fig. 340.

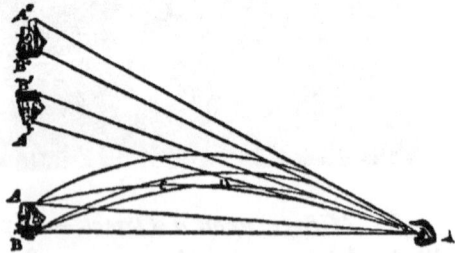

In this case the layers of air near the cold surface of the water are considerably colder and denser than those higher up. Rays, therefore, which pass upward from an object are continually entering rarer layers of air, and are therefore bent more and more downward, as shown in Figure 340. If the rays *A C* and *B D*, coming from the top and bottom of the object, are totally reflected at the points *C* and *D*, they will cross on their way to the eye, and cause the object to appear *elevated and inverted* at *A' B'*. If the rays coming from the top and bottom of the object are simply bent round without being totally reflected, they will not cross before entering the eye, and the object will appear *elevated and erect*, as at *A" B"*. The elevation of an object by refraction without inversion is sometimes called *looming*. Sometimes objects entirely below the horizon are elevated by refraction sufficiently to appear distinctly above the horizon.

448. *The Rainbow.* — The rainbow, when complete, is *a colored arc having a radius of about* 41°, *and containing all the prismatic hues, the red being on the outside and the violet on the inside.* There is often *a second fainter bow, with its colors in the reverse order,* outside of the primary bow. This is called the *secondary* bow. Occasionally, there are one or more *supernumerary* bows within the primary bow, composed of colored arcs of greater or less extent.

The rainbow appears whenever the sun shines upon falling rain in the opposite part of the heavens. The bow is never seen unless the sun is within 41° of the horizon, and the nearer the sun is to the horizon the larger the arc of the bow.

A *line drawn from the sun through the eye of the observer* points to the centre of the circle of which the rainbow is a part, and is called the *axis* of the bow. A line drawn from the eye of the observer to the centre of the colored band at any point makes an angle of about 41° with the axis of the bow. A line drawn from the eye of the observer to the red edge of the bow makes an angle of about 42½° with this axis ; and one drawn to the violet edge, an angle of about 40½°.

The rainbow is produced by *rays of sunlight reflected from the rear surface of the rain-drops.* These rays would be refracted both on entering and leaving the drops. At each refraction they would be bent towards a line drawn to the point of contact of the ray with the rear surface of the drop, and parallel with the incident ray of sunlight, and therefore parallel with the axis of the bow (Figure 341). At an angle of 41° with the axis of the bow the rays emerge from the rain-drop crowded together and *almost parallel* with each other (Figure 342). These rays are able to preserve their intensity through long atmospheric distances. At all other angles the emergent rays are divergent, and become too feeble to affect the eye. Accordingly, whenever the observer looks 41° away from the axis of the bow, his eye catches some of

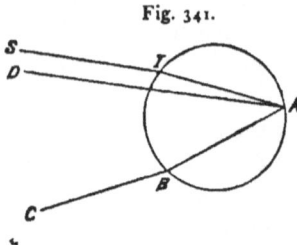

Fig. 341.

Fig. 342.

these nearly parallel rays which are emerging from some rain-drop. He therefore sees a bright band, circular in form, and having a radius of 41°.

The different colored rays are refracted unequally on their passage through the rain-drop; hence the angle of parallelism is somewhat different for different colors, being about 42½° for the red and about 40½° for the violet. This accounts for the colors of the rainbow, the violet rays reaching the eye from drops nearer the axis than those which send red rays to the eye.

No two observers see the same rainbow; that is to say, no two eyes receive the colors from the same set of rain-drops.

Fig. 343.

The *secondary* bow is produced by rays that have suffered *two reflections* within the rain-drops (Figure 343). Figure 344 shows the relative position of the two bows.

Fig. 344.

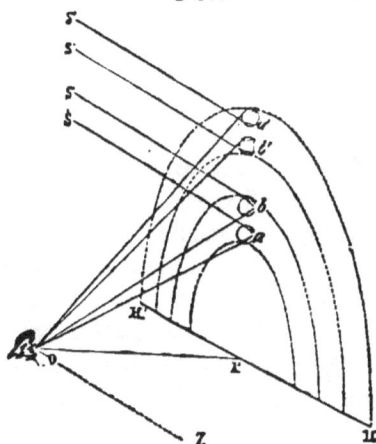

B. Reflection.

449. *Diffused Daylight.* — When the sun shines upon any portion of the atmosphere, *the particles of air reflect the rays of light irregularly*, and so scatter the light in

every direction, thus giving rise to *diffused* daylight. Were
it not for the atmosphere, shadows would be utterly devoid
of light, and rooms into which the sun was not directly
shining would be totally dark.

450. *Twilight.* — Were it not for the atmosphere, the
darkness of midnight would begin the moment the sun
sank below the horizon, and would continue till he rose
again above the horizon in the east, when the darkness of
the night would be suddenly succeeded by the full light of
day. The gradual transition from the light of day to the
darkness of the night, and from the darkness of the night
to the light of day, is called *twilight*, and is due to the
diffusion of light from the upper layers of the atmosphere
after the sun has ceased to shine on the lower layers at
night, or before it has begun to shine upon them in the
morning.

Twilight begins and ends when the sun is about 18°
below the horizon.

451. *Color of the Sky.* — Large particles reflect and diffuse
all luminous waves equally well, but a particle intermediate in
size between a red and a violet wave would reflect a greater
proportion of violet waves than of red waves. The smaller the
particles suspended in a transparent medium, the greater the
proportion of blue rays reflected and the less the proportion of
red. Hence any transparent medium holding very minute particles of any kind in suspension will appear blue in reflected light.
According to Tyndall, the sky owes its blue color to the *minute
particles of watery vapor or other substances suspended in it.*
The more minute the particles, the bluer the sky. As we approach the horizon the sky inclines to white, because of the
larger particles which are present in the lower layers of the
atmosphere.

When the sun is near the horizon, the rays traverse a greater
atmospheric distance, and the separation between the long and
short waves is more complete. In this case the rays which
reach us, and which illumine the clouds and the lower portion of
the sky, are those which are allowed to pass the particles, and

not those which are reflected by them. Hence the evening sky inclines to yellow, orange, or red, according as the shorter waves have been more or less completely turned back. Unless there are clouds in the upper portions of the sky, these colors are limited to the regions near the horizon, since it is there only that the particles in the air are large enough to reflect the larger waves transmitted to them.

C. CORONÆ AND HALOS.

452. *Coronæ.* — When light fleecy clouds pass over the sun or moon, one or more *iris-colored rings* are often seen about these bodies, the inner ring being from 3° to 6° in diameter. The *blue edges* of these rings are *towards the sun or moon*, and the red edges away from it. These rings are called *coronæ*. They are more frequently noticed about the moon than about the sun, owing to the dazzling brilliancy of the latter. They are shown at the centre of the lower part of Figure 345.

Fig. 345.

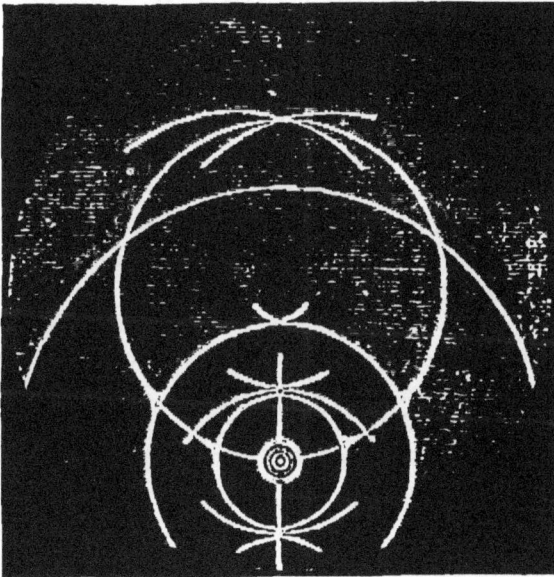

453. *Halos.* — Halos are circles formed around the sun or moon. When bright they are seen to be composed of the pris-

matic colors. Thay are *larger than coronæ*, and are *red on the edge towards the sun*. The halo most often seen has a radius of 22°. This is shown at *h h* (Figure 345). A second halo is sometimes formed having a radius of 46°, *H H*; and occasionally a third halo is seen having a radius of about 90°, *H' H'*.

454. *Parhelic Circle.* — When a halo is formed around the sun we often notice a *white circle passing through the sun* and parallel to the horizon (Figure 345). This is called a *parhelic* circle. It never exhibits prismatic colors like the first-mentioned halos.

455. *Parhelia.* — Near the points where halos cut the parhelic circle there is a double cause of light, and here the illumination is sometimes so great as to present the appearance of a *mock sun, p p* and *P P* (Figure 345), and is called a *parhelion*. Parhelia are generally red on the side which is toward the sun, and they sometimes have a prolongation in the form of a tail several degrees in length, whose direction coincides with that of the parhelic circle.

456. *Contact Arches.* — Arcs of colored circles with variable curvatures are sometimes seen touching the halos of 22° and 46° at their highest and lowest points, *a, b* (Figure 345). Sometimes we notice two arcs of circles nearly white, *A*, intersecting the parhelic circle at a point directly opposite to the sun, and inclined to this circle at angles of about 6J°.

457. *Vertical Columns passing through the Sun.* — Sometimes, near sunset, we notice a luminous column perpendicular to the horizon, rising from the sun to the height of 10° or 15°, and occasionally still higher. Sometimes a little before sunset, a similar column of light is seen to shoot down from the sun toward the horizon. Sometimes columns are seen simultaneously both above and below the sun; and if the halo of 22° is seen at the same time, this column, together with the parhelic circle, presents the appearance of a rectangular cross within the halo (Figure 345).

VIII.

THE THREE GREAT CIRCULATIONS OF THE GLOBE.

458. *The Atmospheric Circulation.* — In the atmospheric circulation, which gives rise to the various systems of winds, masses of air are kept moving round and round. This circulation is maintained by *heat received from the sun, and absorbed by the atmosphere.* The heat thus absorbed causes the air to expand, rise, and overflow, while gravity pulls the colder and heavier air down and around to supply its place. The mechanical energy of the moving masses of air is *exactly equal to the energy of the solar radiations consumed in maintaining the motion.* The energy of the solar radiations absorbed by the air is transformed by expansion into the mechanical energy of the winds. *Winds are merely transmuted sunshine.*

459. *The Aqueous Circulation.* — In the aqueous circulation, water is continually passing into the atmosphere as vapor, then falling from the atmosphere as rain, and, finally, running in various streams down to the level of the ocean. This circulation is also maintained by *energy absorbed from solar radiations.* The solar heat absorbed by water converts it into vapor, and raises it into the atmosphere. When this vapor condenses in the atmosphere, gravity draws it to the surface of the earth, and to the level of the ocean. In the evaporation of the water the kinetic energy of the solar radiations is converted into the potential energy of molecular separation, and in the expansion by which this vapor is raised into the atmosphere, into the potential energy of mechanical separation. In the condensation of the vapor in the atmosphere, its potential energy of molecular separation is transformed into the kinetic energy of heat, and in the fall of the rain to the earth and the descent of the water to the sea, its potential energy of mechanical separation is transformed into the kinetic energy of mechanical motion. The *energy of the mountain stream* which drives the mill came originally to the earth in *the minute vibrations of the solar radiations,* and was absorbed from these by water and air.

460. *The Circulation of Carbon.* — Carbon exists in the at-

mosphere in carbonic acid gas, a compound of carbon and oxygen. This gas is absorbed from the atmosphere by leaves of plants, in which it is decomposed by *solar radiations*, which are also *absorbed by the leaves*. The carbon is retained by the plant, and the oxygen is restored to the atmosphere. When vegetable substances are consumed by the natural process of decay, or as food in the bodies of animals, or as fuel in our stoves and furnaces, the carbon again unites with the oxygen and forms carbonic acid, which passes back into the atmosphere. Thus carbon is kept going round and round, from the atmosphere to plants and animals, and back again into the atmosphere.

This circulation, like the other two, is maintained by *energy obtained from solar radiations*. By the decomposition of the carbonic acid in the leaves of the plant, the kinetic energy of the sunbeam is transformed into the potential energy of chemical separation; and in the consumption of food and fuel, the potential energy thus required by carbon is converted into kinetic energy again. Animals derive all their energy from the food which they eat, and as this food is consumed in the body, its potential energy is converted partly into the kinetic energy of heat, and partly into the kinetic energy of mechanical motion. The *energy employed by man* in thinking, writing, speaking, or in doing any kind of work whatever, came to the earth originally from the sun in *the minute vibrations of the ether*.

Coal is a vegetable substance, and *its potential energy has been derived from solar radiations;* and when we burn coal for fuel or coal gas for light, we are simply extracting from the coal the sunbeams that were ages ago absorbed by the leaves of plants and transformed into the potential energy of chemical separation.

461. *Source of Terrestrial Energy.* — Nearly every form of terrestrial energy is *derived from the sun* and comes to the earth in the solar radiations. The three chief agents for absorbing this energy and transforming it into a kind adapted for our use are *water, air,* and *leaves of plants.*

INDEX.

A.

Aberration chromatic, 170.
    spherical, 170.
Action and reaction, 7.
Affinity, 3, 7.
Air-chamber in pumps, 84.
Air, pressure of, 67.
Air-pump, the, 65.
Ampère's rule, 222.
Anion, the, 232.
Anode, the, 232.
Aqueous circulation, the, 307.
Archimedes's principle, 58.
Artesian wells, 74.
Astatic galvanometer, 223.
    needle, 223.
Astronomy, 4
Atmosphere, circulation in, 307.
    composition of, 257.
    condensation in, 277.
    electricity in, 292.
    height of, 257.
    humidity of, 268.
    movements of, 270.
    reflection in, 303.
    refraction in, 298.
    temperature of, 260.
    weight of, 258.
Atoms, 1.
Aurora, the, 296.
Avogadro's law, 64.

B.

Balance, the, 29.
Balance-wheel, compensation, 117.
Billoons, 60.
Barometer, the, 81, 258.
Beam, defined, 152.
Beats, musical, 103.
Bell's telephone, 237.
Boiling, 127.
Brocken, the spectre of the, 183.
Bunsen's cell, 228.

C.

Calorimeters, 132.
Camera obscura, the, 177.
Capillarity, 76.
Capstan, the, 49.
Carbon, circulation of, 308.
Cathode, the, 232.
Cation, the, 232
Centre of gravity, 23.
Centrifugal force, 9.
Centripetal force, 10.
C. G. S. system, 13.
Charles's law, 65.
Chemistry, 4.
Climates, marine and continental, 265
Clocks, 37.
Clouds, 281.
Cog-wheels, 48.
Cohesion, 3, 53.
Coil, the induction, 240.
Collision of elastic bodies, 16.
Color-blindness, 190.
Color chart, the, 186.
    of the sky, 304.
    scale, 187.
Color-disc, the ideal, 186.
Colors, complementary, 186
    from absorption, 190.
    primary, 187.
Condensation, 129.
Congelation, 124.
Contact-arches, 306.
Coronæ, 305.
Cottrell's straw electroscope, 204.
Cryophorus, the, 135.
Crystals, 90.
Cyclones, 291.

D.

Daniell's cell, 228.
Daylight, diffused, 303.
Density, 6.
Dew, origin of, 277.
Dew-point, the, 269.
Diathermanous bodies, 145.